ENDORSEMENTS

Fascinating, educational, and a great read! Chasing Centuries *fills a large void in our understanding of how humans and agaves have coevolved . . . well done!*
— Tony Avent, Owner of Plant Delights Nursery, and founder of Juniper Level Botanic Garden in Raleigh, North Carolina

Chasing Centuries *is probably about the most comprehensive overview of Arizona agaves likely to be created. The information on distributional ranges of the various species, possible hybrids between them, and ethnobotanical history of the naturalized cultivars formerly utilized by Native Peoples is by far the most thorough I've yet seen. Accompanied by numerous detailed and interesting photographs, this book is a worthy addition to the library collection of any agave enthusiast and lovers of succulents in general.*
— Jan Emming, Creator of Destination: Forever Ranch and Gardens, and owner of Desert Sense Nursery in Yucca, Arizona

This is an exciting time for agave research, involving researchers from all disciplines and interests, including archaeologists, botanists, Native Peoples, and others. This book, with its many fine photographs, helps make the science of this fascinating but complex group more accessible to the public, who can better appreciate and care about these wild and domesticated, unique Arizona gems.
— Wendy C. Hodgson, Curator of the Herbarium and Senior Research Botanist, Desert Botanical Garden, and author/illustrator of *Food Plants of the Sonoran Desert*

Whether you are an agave enthusiast or have a passion for anthropology of the American Southwest, Chasing Centuries *provides an informative and visually stunning synthesis of both disciplines. A longtime devotee of the genus, Parker has done the research and fieldwork, culminating in a beautifully designed book. Gorgeous images of the plants and pre-Columbian ruins are supplemented with maps to paint a thorough picture of agaves and their importance to the native cultures. One quick look-through and I knew it belonged in my library.*
— Jeff Moore, owner of Solana Succulents in Solana Beach, California, and author of *Aloes and Agaves, Soft Succulents,* and *Under the Spell of Succulents*

This book synthesizes the scientific work carried out mostly by a team of researchers at the Desert Botanical Garden in Phoenix, supplemented by the author's own fieldwork in Arizona. The text carefully identifies each of the five indigenous pre-Columbian Native American cultivars that have survived to the present as well as the dozen naturally occurring Arizona Agave species. Hybridization between various taxa is also covered.... After more than five centuries, most of these cultural relics are today threatened with extinction. The author's superb photographs depict every Agave known so far in Arizona.

—Amadeo M. Rea, Research Associate, University of San Diego, and author of *At the Desert's Green Edge*

A must read in the tradition of John McPhee and Edward Abbey. Archaeological-botanist Ron Parker's highly illustrated Chasing Centuries *is a one-of-a-kind travel-adventure-botanical-history-book that properly elevates one of the Southwest's most iconic and life-sustaining plants to its proper place culturally and anthropologically.*

—Stuart Rosebrook, Senior Editor, *True West Magazine*

Chasing Centuries *is an intriguing insight into the world of amazing agaves found in Arizona and is a must read for any agave enthusiast. Parker introduces the reader to long lost cultures of Arizona and their use of agaves in their daily life, and he provides valuable information on the wild agaves of Arizona with excellent distribution maps and knowledge he has learned during his extensive travels into the rugged back-country seeking out these fantastic plants.*

—Greg D. Starr, owner of Starr Nursery in Tucson, Arizona, and author of *Agaves: Living Sculptures for Landscapes and Containers*

This lavishly illustrated and meticulously researched work takes the reader on a fascinating adventure through thousands of years of history of human-agave coevolution in the rugged landscapes of Arizona. Chasing Centuries *is a book to be savored, carried into the field, kept as a reference and gifted to anyone interested in gaining a deeper understanding of how Arizona's ancient peoples played an enduring role in shaping the natural habitats of the region.*

—Michael Wilken-Robertson, anthropologist and author of *Kumeyaay Ethnobotany*

CHASING CENTURIES

Chasing Centuries

THE SEARCH *for* ANCIENT AGAVE CULTIVARS
Across the Desert Southwest

RON PARKER

SUNBELT PUBLICATIONS, INC.

*Chasing Centuries: The Search for Ancient Agave Cultivars
Across the Desert Southwest*

Sunbelt Publications, Inc.
Copyright © 2019 by Ron Parker
All rights reserved. First edition 2019

Cover and book design by Michael Schrauzer
Project management by Deborah Young
Printed in Korea

No part of this book may be reproduced in any form
without permission from the publisher.

Please direct comments and inquiries to:

Sunbelt Publications, Inc.
P.O. Box 191126
San Diego, CA 92159-1126
(619) 258-4911, fax: (619) 258-4916
www.sunbeltpublications.com

22 21 20 19   4 3 2 1

Library of Congress Cataloging-in-Publication Data

Names: Parker, Ron, 1954- author.
Title: Chasing centuries : the search for ancient agave cultivars across the
   desert southwest / by Ron Parker.
Description: First edition. | San Diego, California : Sunbelt Publications,
   Inc., 2019. | Includes bibliographical references and index.
Identifiers: LCCN 2018038148 | ISBN 9781941384480 (softcover : alk. paper)
Subjects: LCSH: Agaves--Desert Southwest. | Agaves--Varieties--Desert Southwest.
Classification: LCC SB317.A2 P28 2018 | DDC 633.5/7709791--
dc23 LC record available at https://lccn.loc.gov/2018038148

All photographs are by the author unless noted or in public domain.

Please note that range maps exclusively represent *Agave* populations
that the author has personally visited and documented.

TABLE OF CONTENTS

Preface ix
Acknowledgments xi
Introduction xiii
Part I: The Historical Perspective 1
   1   Regional Pre-Columbian Cultures 3
   2   Dry Farming Techniques and Ancient Agave Cultivation 17
   3   Traditional Agave Use 25

Part II: Agaves of the Region 29
   4   A Brief Introduction to Agaves 31
   5   Naturally Occurring Agaves 37
   6   Pre-Columbian Agave Cultigens 91

Part III: Notes from the Field 119
   7   The Chase 121
   8   Impact on Naturally Occurring Agaves 131
   9   Cultivation of Naturally Occurring Agaves 137
  10   Inevitable Extinction 143

Part IV: Addenda 151
Table of Agave Species 152
Glossary 153
Endnotes 156
Bibliography 157
Index 160

The rare habitat hybrid, *A. parryi* × *A. schottii*.

PREFACE

I REMEMBER MY FIRST TIME IN AGAVE COUNTRY AS IF it were yesterday. I recall an immediate and overwhelming sensation upon spotting my very first *Agave* in a natural habitat. Of course I had noticed agaves, but gave them very little due while whizzing by along this state route or that on my way to somewhere else. Don't get me wrong, I appreciate those scenes, and am gratified to know they are there, so close to where I had chosen to take up residence. But this was the first time it was personal. We were going to meet face-to-face, and get to know each other. Over the next few minutes I studied leaf shape and tone, spines, both terminal and marginal, new growth, old growth, and the geometry of such a wondrous living organism. I was smitten, lost in awe and wonder.

I went on to find many more *Agaves* that day, along with *Yucca*, *Nolina*, *Dasylirion*, and all manner of interesting cactus. I recall eventually heading back, more than a bit befuddled about what I had experienced and seen. I wasn't sure which agave species were in play. Some appeared to be *Agave chrysantha*, while others resembled *Agave parryi*, and still others seemed like both. Or neither. Turns out my confusion was not without cause. As I came to realize, I had first encountered *Agave chrysantha*, then *Agave parryi*, farther along at higher elevation. In between was a happy, healthy population of intermediates. Of course, I had no appreciation for intermediate or hybrid forms at the time, but have since come to realize that *A. chrysantha* × *A. parryi* hybrids are among the most ubiquitous agaves found across certain parts of central Arizona.

They say all great journeys begin with a single step, and if so, that was mine. A week later, I was up in the mountains again, not far from my first adventure. No *Agave parryi* this time, just lots of *Agave chrysantha*. Each and every new plant was a marvel to be examined and admired. After a while I came across something different. No confusion this time—I knew this was my first habitat encounter with *Agave toumeyana*. So distinct was this diminutive agave, it was nearly inconceivable. Eventually, I returned home and pored over hundreds of photographs. In truth I enjoyed studying my photos nearly as much as the original encounters they captured. And that is still true today.

This continued for some time, as I expanded my habitat horizons to explore more and more terrain. Then it happened. I found something new and unexpected. Bewildered by this unfamiliar agave, I took dozens of photos and resigned myself to investigate further once I returned home. That subsequent investigation led to the realization that I had encountered my very first pre-Columbian cultigen, *Agave delamateri*. As I researched this intriguing agave and its marvelous history and association with mysterious ancient cultures, I was smitten all over again. Over the next few weeks, I found more *Agave delamateri*, as well as nearby ancient habitation sites, which were every bit as thrilling to explore as the agaves. Oh my, what had I stumbled onto?

Well here we are several years later. My initial excursions inspired a healthier lifestyle regimen in order to better endure the rigors of mountain hikes. I had also procured a couple off-road vehicles to get as far up into the mountains as possible before embarking upon those hikes. Along the way, I've met and compared notes with others bitten by the botany and/or archaeology bug. I've maintained comprehensive records, documenting everything we've found with routes, waypoints, field notes, floral data, and photographs. Lots and lots of photographs.

Over the past several years, my habitat companions and I have discovered very rare agaves, teetering on the edge of extinction, and nearly untold numbers of pre-Columbian cultigens at more than 300 sites across the state, many of which were previously unknown, along with their associated archaeological sites. And more. Much more. There is still a world of mystery out there, just waiting for someone to come along and pull the right string to have it all unravel in a blinding revelation of insight: the true relationships and evolution of agaves along their odyssey from central Mexico to Arizona, and the role indigenous Native Americans played toward that end. There is so much we don't know, and may never know. Which is fine, it's all about the journey, after all. And the journey continues.

ACKNOWLEDGMENTS

I WOULD FIRST AND FOREMOST LIKE TO THANK ANYone and everyone who has ever accompanied the author into the Arizona wilds for providing extra sets of eyes and for sharing the passion that led us to visit so many challenging and remote sites across the montane regions of Arizona, braving all manner of weather, wildlife, rednecks, crazy drivers (*ahem*), and terrain. Foremost among these habitat companions is a longtime resident of central Arizona named Peter Flitner. His keen eye, knowledge of regional flora, and appetite for adventure proved extremely valuable on many an excursion.

I'll also shout out to Greg Starr of Starr Nursery in Tucson, AZ, who vested in me a strong sense and understanding of the underpinnings of agave taxonomy. I brought the lessons of his superb hands-on instruction along on every excursion. Of course, Greg also provided his own brand of wit, humor, and perspective on many an excursion.

I would like to thank the fine agave research team from the Desert Botanical Garden in Phoenix. Wendy Hodgson, the renowned botanist, researcher, and author of formal descriptions for four of our five domesticated agaves responded carefully and thoughtfully to nearly each and every query I tossed her way over a period of years. I cannot count the times in which her insightful correspondence contributed mightily toward a welcome sense of clarity that occasionally overwhelmed my latest and seemingly impenetrable state of confusion. The lessons learned from working with her in the field provided a stellar example from a consummate professional, of admirable discipline and methodology to emulate.

And finally, thank you to Andrew Salywon, also of the Desert Botanical Garden, for his astute insights and observations while out chasing agaves in the field, and for his important work in regard to establishing affinities and correlatives of our beloved agaves via DNA analysis.

Welcome to the wilds of Arizona.

INTRODUCTION

Dotting the montane wilds of our Arizona landscape are all manner of treasured archaeological site, long since abandoned by prehistoric residents of six ancient cultures: Ancestral Puebloan, Hohokam, Mogollon, Patayan, Salado, and Sinagua, along with an interesting assortment of rather unusual agaves with which some are apparently associated. These agaves appear to be anthropogenic cultigens (see Glossary); living archaeological relics developed and planted by indigenous pre-Columbian Native Americans, and many are still growing exactly where they were planted hundreds of years ago.

Agaves are succulent New World monocots, comprising more than 200 species with an epicenter in Mexico. They also extend into Central America, South America, the Caribbean, and the US, occupying habitat in FL, TX, NM, AZ, UT, NV, and CA, with the number of taxa in AZ exceeding the other states combined. There are currently 22 recognized taxa in Arizona, most natural, but some of apparent anthropogenic origin. There are also many agaves that don't fit neatly under these taxonomic constraints. Some are intermediate, others may be compromised by feral pre-Columbian cultigens, and still others are indeterminate for reason or reasons as yet undiscovered. I am regularly confronted by these types of mysteries as I continue to explore, and wouldn't have it any other way.

A detailed years-long survey of agave populations across Arizona has led to some startling observations and revelations, as that field experience suggests that the very best perspective from which to examine the distribution and unique qualities of many regional agaves is through a lens of pre-Columbian cultigen. These fascinating agaves, which I had at first believed isolated and peripheral, actually seem to be everywhere, often right under our noses, hiding in plain sight.

What follows is a brief introduction to ancient cultures associated with pre-Columbian agave cultigens, the dry farming techniques they employed, and their reliance upon agave as a critical natural resource. I'll provide an overview of naturally occurring agaves in the region within a context of the cultigens they crafted and chose to use in their stead, and explore the relationship between these ancient cultures and the succulent horticultural charges they left behind for us to admire. I'll consider the who, what, where, when, and why, as well as the influence these cultigens continue to wield over naturally occurring agave populations today.

A striking Sinagua pictograph panel from red rock country.

PART I
The Historical Perspective

Archaic petroglyphs from west-central Arizona may be 3,500 years old.

 CHAPTER 1

Regional Pre-Columbian Cultures

THE EARLIEST HUMAN HABITATION IN THE AMERI-can Southwest dates to roughly 16,000–20,000 years ago.[1] Paleo-Indians were among the first inhabitants of the New World, and utilized habitat near fresh water sources, which had an abundance of fish and attracted birds and game animals. Big game, including bison, ground sloths, and mammoths were also attracted to these areas.

As climate warmed at the end of the Ice Age, megafauna began to disappear. Hunter/gatherers gradually adapted to these changes, supplementing their diet with a variety of plant foods and smaller game. Archaic people used various tools to hunt water fowl, deer, and smaller mammals. Hunting was especially important during winter.

The nutritive value of grass and other seed was exploited as flat rocks were used to grind flour to produce gruels and breads, and this marks the beginning of what archaeologists call the Archaic Tradition. The Archaic period is dated from roughly 8,000 to 2,000 years ago, and encompasses the lengthy span that separates the end of big game hunters and the rise of agricultural societies. During this time people of the southwest developed a variety of successful survival strategies. Small bands traveled through the area, gathering food such as cactus fruits, mesquite pods, acorns, and pine nuts. Archaic people lived in the open, but may have built temporary shelters within camps at collection points, and they returned to these places year after year. Archaic petroglyphs are still found today at some of these ancient encampment sites.

The post-Archaic period begins at roughly AD 200, and agaves may have had as much to do with the gradual shift from hunter/gatherer to agrarian culture as any other plant group.[2] Agaves were a critical resource for ancient indigenous Americans, though not the most convenient fare to transport back to encampment sites. Although plentiful in many areas, the natural range of agaves is generally limited to foothills and low to mid mountainous regions. In many instances, naturally occurring agaves did not grow alongside habitation sites, but rather at some lateral distance, or perhaps 200–400 m (650′–1,300′) above.

Since many agaves reproduce asexually via rhizome, offsets undoubtedly provided convenient fare to uproot and transplant closer to home, once that spark of initiative saw the light of day. And it would not have taken long to learn to do so at a time coinciding with the availability of seasonal moisture in order to successfully establish offsets in new gardens, with supplemental water applied as needed. Classic agricultural basics, such as collecting and sowing seed, from agaves and other valued plants, probably came later in the game.

Several prominent ancient Native American cultures developed during this period, some of which are strongly associated with unique pre-Columbian agave cultigens still found in Arizona today. Our story continues with a brief introduction to each.

Archaeological cultures associated with pre-Columbian agave cultigens.

Hohokam kiva found near *A. murpheyi* dates to the Sedentary Period, roughly AD 900–1150. Perimeter walls are close to 4 m tall!

The HOHOKAM (from the O'odham word *Huhugam*, meaning "those who are gone") was a pre-Columbian Native American culture that colonized central Arizona from roughly AD 450–1450.[3] It is the oldest such culture in Arizona, associated with specific agave cultigens. Its epicenter was in and around the downtown Phoenix area, but Hohokam territory was expansive, occupying roughly 78,000 km² (30,000 mile²) north through the New River Mountains, and southeast to the Tucson Basin and beyond.

This culture endured for more than a millennium, and of course, evolved substantially across that time. They lived in adobe houses and compounds, some in small rural communities, others in expansive urban settings. The Hohokam were accomplished farmers with an extensive toolbox of strategies from which to draw, designed to cope with hardships of growing crops in the desert, eventually creating a massive network of hundreds of kilometers of canals to irrigate a vast area as large as 70,000 acres. The engineering, construction, and maintenance of this irrigation network would have consumed enormous labor and organizational resources, and likely served as a framework for successful social and economic systems.

Remnant early pit house compounds, later pueblo compounds, earthen ball courts, platform mounds, and petroglyph panels are still found throughout the region, though many of their more populous living centers have been decimated by modern urban development.

Fortified hilltop Hohokam habitation site.

Ball courts and platform mounds are unusual in the desert southwest and suggest a Mesoamerican connection. General aesthetics and architecture also suggest a general connection to some well-known Mesoamerican cultures, including the likes of Aztec, Maya, and Toltec. Well-established trade routes and artifacts such as figurines, shell jewelry, and pottery suggest an even more direct connection to western Mexico.

Although evidence of agave cultivation dates back through the very inception of Hohokam culture, large scale agave production does not appear to have commenced until at least AD 950,[4] and it does not seem unlikely that advantages provided by a pre-Columbian cultigen may have contributed to this shift.

Our most meaningful encounters with Hohokam archaeological sites are around the northern periphery rather than major population centers, since that periphery is currently less populous and at higher elevation, thus more conducive toward the survival of agaves that once graced long-abandoned gardens and farms. A single pre-Columbian cultigen, *A. murpheyi*, is associated with these Hohokam archaeological sites.

The Casa Grande Ruins National Monument in Coolidge, AZ. and the Pueblo Grande Museum in Phoenix, AZ, are terrific destinations for a modern-day hands-on Hohokam experience, as is the Painted Rock Petroglyph Site in Dateland, AZ.

Hohokam petroglyphs associated with a couple of *A. murpheyi* stands.

Very busy Hohokam petroglyph panel.

Hohokam metate used for grinding corn.

The SINAGUA (Spanish: without water) represent an enigmatic culture that may have arisen from Yuman origins, or possibly an offshoot of Mogollon culture from eastern AZ and NM. Other archaeologists suspect a regional culture influenced by neighboring Ancestral Puebloan (aka Anasazi, a Navajo term meaning "ancient enemy," which many modern Puebloans find offensive), and Hohokam. Regardless of origin, the Sinagua occupied lands from north of Flagstaff to the Verde Valley from roughly AD 650–1350. Archaeologists like to divide the Sinagua into northern Ancestral Puebloan-influenced and southern Hohokam-influenced sects. Our interest is with the Southern Sinagua, who eventually displaced Hohokam in the extreme southern part of their range.

A rich cultural infusion after AD 1085[5] may have resulted from cataclysmic volcanic activity in the area known as Sunset Crater. This series of massive seismic events dramatically changed Sinagua culture, and attracted distant members of other religious cultures to the region to see firsthand what wrath the gods had wrought. Volcanic activity in many ways must have defined the very existence of these people, and continued sporadically for nearly 150 years.

Sinagua cliff dwelling from the Honanki Phase associated with several domesticated agave stands.

Little known Sinagua cliff dwelling from roughly AD 1200 at 1,500 m elevation.

Although culturally distinct, Sinagua appear to have adopted farming techniques from Hohokam, as well as building techniques from Ancestral Puebloan, and probably traded regularly with both. In fact, the farther south one travels through Sinagua territory and the Verde Valley in particular, the more Hohokam-like the ancient dwellings and artifacts. In addition to adobe compounds similar to those crafted and occupied by Hohokam, later-period Sinagua resided in cliff dwellings along canyon walls in and around red rock country near Sedona. Remnants of many such dwellings and exquisite rock art panels are still found in the area today.

Our most meaningful encounters in Sinagua territory start north of Sedona and extend south through the entire breadth of the Verde Valley. Four formally described pre-Columbian agave cultigens reside here, two of which appear to be endemic to the area.

For a personal glimpse of breathtaking Southern Sinagua archaeological sites, plan a visit to the Palatki Heritage Site in Sedona, AZ, and the V Bar V Ranch and Montezuma Well in Rimrock, AZ. Montezuma Castle National Monument in Camp Verde, AZ, and Tuzigoot National Monument in Clarkdale, AZ, are also open to visitors year round. Wupatki National Monument in Flagstaff, AZ, offers a phenomenal experience and easy access to dozens of Northern Sinagua archeological sites.

This fully intact dwelling from the extreme southern edge of Sinagua territory dates to roughly AD 1200, and now plays host to large numbers of Mexican free-tailed bats (*Tadarida brasiliensis*).

Pictograph panel from red rock country north of Sedona.

Sinagua petroglyphs found north of Sacred Mountain, in close proximity to *A. verdensis*.

A nearly intact Salado pueblo found inside a cave near the Superstition M...

The **SALADO** (Spanish: salty, a reference to the Salt River) culture supplanted that of the Hohokam through the eastern part of its range from roughly AD 1150–1450.[6] The term "Salado" is actually a tad nebulous and may still be evolving in regard to the accepted extent of its range. The area central to the Salado had been sparsely settled by outlying Hohokam and others long before, but an ensuing population influx eventually redefined the region. The Salado essentially represent a merger of Hohokam and migrating Mogollon-influenced Ancestral Puebloan cultures, with its major population center in the Tonto Creek area. The Globe/Miami area, San Pedro Watershed, and Safford Basin also took form as cultural melting pots in which diverse peoples forged alliances and attended to the business of farming.

Some archaeologists argue whether the Salado culture merits the same degree of recognition as the Hohokam. There are important contrasts; neither Ancestral Puebloan kivas nor Hohokam ball courts are found near Tonto Creek, and this suggests distinct religion, thus distinct culture. From an agricultural, and more specifically an Agavaceae standpoint, the Salado also seem at least marginally distinct.

Unfortunately from an historic perspective, a large 40 km (25 mile) long tract of land at the center of Salado territory was dramatically flooded more than 100 years ago by the construction of the Roosevelt Dam, inundating

Remains of one of several large Salado pueblos at Sierra Ancha.

Three-story Salado pueblo, affectionately called "The Crack House" found within the Sierra Ancha Wilderness.

Large Salado dwelling, just north of Roosevelt Lake, near several large *A. delamateri* stands.

untold numbers of archeological sites. Prolonged dry spells occasionally lower water levels at Roosevelt Lake to expose the remains of submerged prehistoric villages around its perimeter.

Although rock art is not broadly associated with the Salado, platform mounds, hilltop pueblos, and several extraordinary cliff dwellings are found in Sierra Ancha and the Tonto Creek region. Sierra Ancha cliff dwellings are remote and challenging to reach, but are among the most impressive archaeological sites in the state.

Our most meaningful encounters in Salado territory are those north of Roosevelt Lake. The area is home to *A. chrysantha*, *A. parryi,* and *A. toumeyana*, and though three different pre-Columbian cultigens also inhabit these haunts, *A. delamateri* absolutely dominates the cultigen landscape, and in numbers that suggest the area may have once been heavily populated.

Tonto National Monument offers a wonderful glimpse of 700-year-old Salado cliff dwellings. The Upper Cliff Dwelling is accessible by guided tour only, and well worth the added effort of making a reservation. Tours are offered November thru April, as weather permits. Besh Ba Gowah Archaeological Park and Museum in Globe, AZ, offers visitors a journey through time to a partially restored Salado pueblo.

END OF AN ERA

It is a matter of great historical and archaeological interest that these regional Native American cultures appear to have simply vanished from the face of the planet sometime around AD 1450–1500. It was not until 1540 that Francisco Vasquez de Coronado marched north into Arizona and first met regional Native Americans, and by that time the heroes of our saga were long gone. Tree ring data suggests climactic upheaval and severe flooding impacted the region from AD 1400–1450.[7] Additional theories abound regarding prolonged drought, salted fields, political instability, disease, conflict, and migration, but in the end there are few clues from which to draw. Some of our best clues may come from contemporary Native American folklore, which suggest that Sinagua migrated east and are now part of the Hopi and Zuni tribes. Other tales suggest ancient Hohokam may have migrated north, or even been subsumed by the Tohono O'odham Nation. I know of no tribal legends implicating current whereabouts of the Salado.

CHAPTER 2

Dry Farming Techniques and Ancient Agave Cultivation

ANCIENT NATIVE AMERICANS EMPLOYED DIFFERENT strategies for coping with harsh realities of growing crops in the desert. The most basic and fundamental of these involved planting those crops best suited for arid conditions, then further refining and enhancing that suitability by utilizing selection and (possibly) hybridization techniques. Selection is a process of refinement in which those examples that best demonstrate desirable traits are specifically chosen to provide genetic material for succeeding generations, and those with undesirable traits are summarily eliminated.

Consider maize, as an example. It has only recently been confirmed through DNA analysis that maize/corn has its origin in teosinte, a tall, naturally occurring Mesoamerican grass. I have on several occasions found corn cobs among the ruins at ancient habitation sites. These cobs are typically 70–80 mm (2.5"–3") long and perhaps 700 years old. Archaeologists have found many examples half that size at older sites, and 5,000-year-old cobs tend to

Eight hundred year old corn cob checks in at about 70 mm.

check in at about 20 mm (0.75"). Compare that with modern corn found in today's food marts as an example of the power of selection.

There is little doubt that ancient indigenous Native Americans have been selecting agaves for certain qualities for a very long time, going back at least as far as the Aztecs and Mayans. *A. fourcroydes* was distinct and fully domesticated by the time Spanish conquistadors arrived, per early reports that henequen grown locally in gardens was of much higher quality than habitat stock from which it was presumably derived.

Beyond such heady horticultural pursuits were a variety of successful hands-on farming techniques. One widely employed strategy was terracing, which involves the transformation of regular or uneven slopes into steps to

Rows of parallel rocks and shallow ridges reveal ancient Hohokam terracing at South Mountain.

maximize arable land. Terracing was often used in conjunction with check dams to regulate and direct runoff to mitigate soil erosion and frost damage. This labor intensive task was managed with extremely primitive digging tools and required regular maintenance, since slopes would otherwise return to some semblance of their original form. Much of the terracing associated with pre-Columbian agave cultigens in Salado and Sinagua haunts is found near permanent waterways. I occasionally find cultigens planted along unusually steep slopes, in which long abandoned terraces had reverted to their natural state.

While terracing was applied to good effect along hillsides and other moderate slopes, different strategies were often required for the flats below. The most widely used and effective such strategy may have been rock or gravel mulching, a wildly successful scheme that desert gardeners can still utilize to good effect today. The idea is to build rock piles directly adjacent to our cultivated desert plants in order to maximize moisture retention and reduce rodent damage. Rainfall easily penetrates rock piles, which dramatically slow evaporation and desiccation. Typical rock piles are variable, but might be 100 cm (40″) across and 50 cm (20″) tall.

In many areas, indigenous Native Americans practiced floodwater farming, which amounts to planting crops in the floodplains of rivers that overflow their banks after major storms. This method was most effective when floods came at predictable intervals. Where floodplains were not apparent, check dams were built or crops were planted in the mouths of arroyos.

Then of course there is the remarkable canal system engineered by the ever industrious Hohokam. Some of the main canals were enormous, 3.5 m (12′) deep and 15 m (50′) across at the bottom, and many were tapered to manage water flow across distances as great as 30 km (18 miles). This vast network irrigated some 70,000 acres of desert farmland, and allowed Hohokam to farm beans, squash, corn, and cotton while agave, more tolerant of arid conditions than other money crops, was relegated to aforementioned dry farming techniques. Rock mulching was not employed in irrigated fields.

A particularly compelling archaeological study organized by Suzanne K. and Paul R. Fish in the early 1980s uncovered evidence of large-scale farming along desert bajada slopes west of the Tortolita Mountains north of Tucson in the general vicinity of Marana.[8] Remains of terraces and check dams are found in association with massive numbers of rock piles. Analysis of charred remains in the area suggests these fields were used exclusively to cultivate agave. The authors conservatively conclude that the 1,200-acre study area was home to more than 100,000 agaves growing together in an ancient Hohokam rock

Dry Farming Techniques and Ancient Agave Cultivation 21

Aerial shot of the Safford grid fields. *Photo courtesy of Dr. James A. Neely.*

mulch farm, which might have produced an annual yield of more than 10,000 harvested agaves, equating to forty metric tons of edible product. A further estimate of fifty man-years is suggested as an initial expenditure in constructing this dry farming field. The rock pile farm dates back to the Hohokam Classic Period, AD 1150–1350. *A. murpheyi* is strongly suspected as the dominant crop species, given the enormous numbers of agave and relatively low elevation, but charred remains offer no clue in that regard. Growing agaves was serious business indeed.

This study was a groundbreaking effort. Ancient agave cultivation had long been established in Mesoamerica, but this was the first solid evidence establishing similar patterns in the southwestern desert. Since that time, more than 550 ancient agave cultivation sites have been documented across central Arizona, an amount significantly greater than the number of sites at which living remnants of ancient agave crops are found.

Then we have the Safford Valley grids,[9] where rock piles still form a discernable pattern, easily seen via aerial recognizance. Grid fields sporadically blanket an area of nearly four square miles, consisting of roughly thirty irregularly

Grid field ground shot shows rock mulch to good effect

shaped and subdivided grids. More than fifty miles of rock alignments provide a farming area in excess of 200 acres.

This is an unusual area, under a cultural umbrella of Hohokam until the late twelfth century, when an influx of Ancestral Puebloan migrants driven from drought-stricken lands tilted the balance toward the enigmatic culture known as Salado. Mimbres and Mogollon artifacts are also found across the region. Archaeologists and ethnobotanists believe a variety of crops were grown here, but a large agave population was central to the ancient rock pile farm. The most likely Agavaceae candidate for this large-scale enterprise is once again, the pre-Columbian cultigen, *A. murpheyi*, now commonly known as the Hohokam agave.

The surrounding region is rife with ancient settlements, terraces, check dams, and irrigation canals, with evidence suggesting the area was farmed as long ago as AD 150. Comprehensive analysis of these sites includes the identification of rock piles, terraces, and roasting pits; radiocarbon dating; and intense scrutiny of charred plant remains, pollen counts, and agricultural artifacts. Conclusions from this exhaustive analysis suggest that the Safford Valley grid site generated approximately one half the agave yield as that of the Marana site, with only 17% as much arable land. This tremendous leap in efficiency might result from improved methodology and more favorable natural conditions.

CHAPTER 3

Traditional Agave Use

INDIGENOUS NATIVE AMERICANS HAVE BEEN HAPPILY coexisting with and utilizing agave for a very long time. Coprolite analysis suggests early Paleo-Indians have been ingesting agave for more than 9,000 years.[10] Additional evidence includes quids and ancient artifacts made from agave fiber. Quids are fibrous remnants of plant material, which are chewed, but not usually consumed, such as chewing tobacco. Large numbers of agave quids have been unearthed at very old archaeological sites in southern Mexico, and these quids have been studied and linked to a particular type of dental wear.[11]

So one prominent use of agave was as a food source. Agaves undergo a dramatic transformation before they produce an inflorescence and bloom. Leaves thin, marginal spines wane, and the base begins to swell as sugars are collected for the production of nectar. This is when agaves are harvested. Leaves are trimmed off, and agave *cabezas* (Spanish for "heads"), now looking for all the world like giant pineapples, are collected and tossed into a makeshift oven for cooking and processing. A pit is dug, and a coal fire started at the bottom. Once a fire is raging, *cabezas* are added, then covered, and another fire started on top. Two to four days later, roasted agave *cabezas* are collected and eagerly consumed. Roasted flesh is soft and sweet, often compared to molasses and sweet potatoes. In addition to consuming product as it came from earthen ovens, a considerable portion might have been worked into cakes, which were thoroughly dried and stored for winter consumption or trade. Flowers and unopened buds could also be processed as food items.

Agaves have also been used for the production of pulque, a fermented white, viscous beer-like drink made from aguamiel, the juice drawn from enormous agaves known as maguey with the use of taps.[12] Large maguey can produce as much as two liters of aguamiel per day for up to a month. Aztec priests associated pulque with the maguey goddess Mayahuel, and considered it extremely sacred, largely restricting its use to religious festivals and ritual sacrifice. The popularity of pulque declined somewhat after Europeans introduced distillation techniques that led to the production of mescal, but pulque still represents roughly 10% of alcoholic beverage consumption in Mexico. Fermented beverage production may have been less prevalent in the desert southwest, since regional agaves do not yield an abundance of sap as do some of their larger Mesoamerican cousins.

In addition to consumables, agave leaves have long been processed for fiber, and terminal spines utilized as needles to craft snares, nets, rope, clothing, pottery, basketry, blankets, and sandals.[13] Two Mesoamerican species in particular, *A. fourcroydes* and *A. sisalana*, considered anthropogenic cultigens, have a proud history of widespread plantation cultivation in Mexico, the US, and abroad toward an end of henequen and sisal production.

Terminal spines have also long been used as construction nails, and one agave part or another crafted as fish stringer, armor, lance, paint, and ceremonial objects. Some agaves have even been used as a source of poison for hunting and fishing. Stalks were used as all manner of building material, make fine walking sticks, and are to this day fashioned into stringed musical instruments and flutes by modern-day Native Americans, many of whom continue to place a

Preparations being made for an agave pit roast, held by a regional archaeology club. Photo courtesy of Scott Newth.

Traditional Agave Use

high premium on trappings of traditional culture. Agaves have also been used as a source of paper since the time of the Mayans.

We know agave syrup boasts certain medicinal qualities. Used as soap or shampoo, it contains bacterial polysaccharides, and saponins and sapogenins, which have antibiotic and fungicidal properties, and may have also been utilized medicinally by ancient Native Americans. The root stock of certain agaves are used as soap and shampoo, and suitable plants (agave and otherwise) are called amole. The soaptree yucca (*Yucca elata*), a close agave relative, derives its common name from this attribute and has been widely used in this manner for centuries.

Modern-day agave use extends toward commercial sweeteners, an enormously successful tequila industry, and the ongoing extraction and development of a very large number of steroidal saponin compounds, some of which have been exploited medicinally for gastrointestinal, anti-inflammatory, and antimicrobial use. Other agave-based steroids have seen time as antitumor and contraceptive agents.[14] Modern agave use also includes livestock feed, living fence construction, and shanty roof shingles. Of course, it would be entirely remiss to overlook the widespread distribution and cultivation of agave as prized ornamental garden and nursery fare.

As mystical and magical as it may seem that a single plant could possibly fulfill so many diverse needs, it is important to note that agaves are a chemically diverse group, and that different species are utilized toward different ends. Agaves with high starch and sugar content, such as *A. palmeri* and our pre-Columbian cultigens, are most fit for consumption. Others with high sapogenin content, such as *A. schottii* and *A. toumeyana*, are most fit for medicinal use, but less so for consumption. And some agaves, such as *A. fourcroydes* and *A. sisalana*, are famously more suited toward fiber production than either consumption or medicinal use. So you'll want a few different agave species to cover all your bases.

Ancient Hohokam petroglyphs found just east of Oro Valley.

PART II
Agaves
of the
Region

Agave blooms grace the landscape east of Roosevelt Lake.

CHAPTER 4

A Brief Introduction to Agaves

BEFORE EXAMINING OUR REGIONAL AGAVES IN detail, it seems advantageous to delve into a few basic principles of morphology, taxonomy, and genetics, since these are the very tools by which we distinguish them. This can prove a more challenging exercise than some might expect, since agaves tend to merge and evolve along morphological gradients rather than the concise delineation suggested by taxonomic decree. There are excellent scientific texts available, which broach these subjects in staggering detail, but the cursory examination offered here is primarily intended to orient readers toward a few basic nuggets of methodology by which field work and observation might translate toward insightful taxonomic conclusion. Also, this brief discussion employs graphic examples that go some small distance toward deciphering haughty taxonomic lexicon broadly employed in our ensuing agave species discussion.

MORPHOLOGY AND DATA COLLECTION

Agaves are succulent New World monocots, comprising more than 200 species with an epicenter in Mexico. They range from very small (10 cm [4"]) to very large (5 m [16']), and form striking spiral rosettes, generously equipped with all manner of sharp tooth and stiff spine to ably protect water stores and deter large foragers. Most agaves are plants of the mid to high desert, usually found at elevations ranging from 900–2,200 m (3,000'–7,200'). Agaves are commonly called "century plants", based upon an old wives' tale that they bloom every 100 years. With few exceptions, these plants are monocarpic (flower only once during their lifetime), but typically live for 8–40 years, rather than 100, before producing an inflorescence in a dramatic flourish of nearly unbridled growth and color. The average lifespan of agaves in Arizona is 15–20 years, give or take.

Agaves are separated into two groups, based exclusively upon the form of the inflorescence. Many agaves, including all of our Arizona pre-Columbian cultigens, feature big broad leaves and a large, elaborately branched inflorescence, referred to as paniculate (arranged in panicles). This paniculate inflorescence can quickly grow as large as a mid-sized tree. These are placed in the subgenus *Agave*. Other agaves generate stalks that are spicate (unbranched) or racemose (intermediate), rather than paniculate, and are placed in subgenus *Littaea*. Spicate stalks of *Littaea* agaves are typically smaller and develop more quickly than those of *Agave* agaves. Intermediate racemose stalks are a frequent result of *Agave* × *Littaea* introgression. Reproductive tendencies vary somewhat from species to species, but agaves generally reproduce both sexually via seed and asexually via rhizome and/or bulbil, small plantlets that form along bloom stalks.

The master of all things agave forever and ever was a man named Howard Scott Gentry (1903–1993), and he devoted a substantial portion of his life toward establishing definitive criteria and methodology for the taxonomy of agaves. The culmination of that work, *Agaves of Continental North America* was published in 1982. It is a monumental tome that any and all true agave lovers should procure and cherish. What Gentry did in a sense seems simple and obvious, but in practice proved exceedingly difficult. He described agave taxa on the primary basis of flower structure, and took vegetative and other characteristics into consideration as secondary factors. The challenge, above and beyond enduring the rigors of chasing agaves across poorly or undeveloped montane regions of Mexico through the mid-twentieth century, was that agaves are monocarpic, and only bloom once toward the end of a modestly long life. This effort took a lifetime of devoted fieldwork and study.

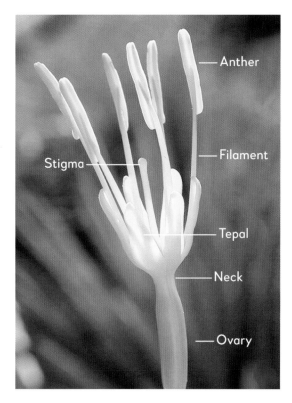

Agave flower, highlighting key components.

Since we all know what a picture is worth, a graphic example of an agave flower is provided, with relevant parts labelled. The notion here is to bisect collected flowers and record measurements of enough examples to produce a range representative of each species. In addition to objective measurements, other more subjective characteristics are also recorded such as color, shape, smell, and firmness. All floral characteristics come to the forefront when describing agave taxa.

A Brief Introduction to Agaves 33

Agave inflorescence comparison: spicate (left) vs. racemose (center) vs. paniculate (right).

    Of course, all manner of vegetative characteristics are also carefully considered, measured, and recorded. This includes plant form, height, and width; number of leaves; leaf shape, length, and width; and inflorescence characteristics including form, height, and number and length of lateral branches (if paniculate). Terminal spine length and decurrence, and marginal spine characteristics are also considered. Finally, GPS coordinates and elevation are recorded, along with the nature of terrain, substrate composition, and other plants in the area. Floral specimens and a good healthy leaf or two are removed and painstakingly prepared as herbarium specimens to accompany recorded data.

    Flower tepal characteristics are of particularly keen interest when considering pre-Columbian cultigens within a context of naturally occurring agaves in the region. Tepals tend to speak broadly to the type of pollinators each agave

species is trying to attract by shaping the business end of flowers toward favorable access for certain pollinators over others. Natural selection is the mechanism employed, since the most successful pollinators generate the largest number of individuals in successive generations.

All of our cultigens feature strictly erect, rather than spreading, tepals. The effect of erect tepals is to contain nectar in quantities large enough to appeal to more robust pollinators, such as hummingbirds and nectar feeding bats. Spreading tepals provide broader access to smaller quantities of nectar, an ideal circumstance for insect pollinators. In addition to erect tepals, four of our five cultigens sport prominent tepal callouses, small brown or reddish tips at the end of each tepal. This is an unusual characteristic, unique to *A. palmeri* among naturally occurring agave species in the region. Tepal callouses are an interesting adaptation that apparently leave nectar slightly more accessible to certain airborne nocturnal pollinators.

Agave flower comparison: Spreading tepals of *A. deserti* var. *simplex* (top) vs. erect tepals of *A. chrysantha* (middle) vs. erect tepals with prominent callouses of *A. palmeri* (bottom).

PLOIDY

Taxonomists rely heavily upon flower and vegetative morphology to appraise and classify agaves, but DNA analysis has also proven a useful addition to the taxonomy toolbox. When agaves reproduce sexually, both seed and pollen donors use meiosis, a special process of division that produces haploid cells (gametes), which contain one half the genetic information of each parent. The object is to meet compatible gametes from another donor in order to reproduce. The success or failure of such meetings lead us to consider ploidy.

Ploidy is a rather complex issue for many different plant types, but is, by the grace of Gaia, fairly straightforward in reference to our small microcosm of the botanic landscape. Ploidy refers to the number of homologous chromosome sets present in a cell, or in the cells of an organism. Examples include monoploid (1 set or X), diploid (2 sets or 2X), triploid (3 sets or 3X), tetraploid (4 sets or 4X), pentaploid (5 sets or 5X), hexaploid (6 sets or 6X), etc. The less specific term polyploid is used to describe cells with three or more chromosome sets, and is particularly useful for describing species with variable ploidy. The term haploid refers to the number of chromosomes within gamete (sexual reproductive) cells. This haploid number represents the number of chromosomes contributed by each partner toward sexual reproduction.

There is a simple shorthand available to intuitively convey these issues. The majority of regional agaves (and humans, for that matter) carry two homologous chromosome sets, thus are diploid, and carry a total of 60 complete chromosomes (humans carry 46, or 23 pairs). This is expressed as $2X$ (diploid) $=60$ (number of chromosomes). Thus 30 is the haploid number, or the number of chromosomes contributed by each gamete in a sexual pairing, and this is expressed as $X=30$. Reproduction of two diploid agaves is thus expressed as $X=30 + X=30 = 2X=60$. So long as an agave is diploid, it can easily hybridize with any and all other diploid agaves.

Some agaves are tetraploid, and carry twice this number of chromosomes. That is expressed as $4X=120$. In this instance, hybridization with diploid agaves is a bit more complicated, but entirely plausible, since we're still working with even factors of the haploid number 30. So here we have $X=30 + 2X=60 = 3X=90$. It works because the tetraploid parent contributes twice as much DNA as the diploid parent to produce triploid progeny. The only issue is that this triploid progeny is effectively sterile, since 90 divided by the haploid number 30 is not an even number.

Of course, chromosomes do funny things sometimes that add considerably to the complexity of these issues, but simple arithmetic adequately covers our bases in the ploidy department most of the time, at least in regard to our regional agaves.

Per Wendy Hodgson and Andrew Salywon of the Desert Botanical Garden in Phoenix, our pre-Columbian cultigens, *A. murpheyi*, *A. verdensis*, and *A. yavapaiensis* are diploid, while *A. delamateri* and *A. phillipsiana* are tetraploid.

BIOINFORMATICS

Agave ploidy has been recorded since at least 1944, so this is hardly groundbreaking stuff, but the relatively new discipline of bioinformatics has the potential to dramatically alter the taxonomic landscape. Bioinformatics is the science of computer-aided analysis of complex biological data such as genetic codes. It can also be referred to as computational molecular biology.

The Human Genome Project declared completion in 2003 after 13 years at an estimated cost of three billion dollars. Work on interpretation and analysis of genome data is still in its early stages. It is anticipated that ongoing analysis will provide the means to achieve significant advances in medicine and biotechnology for many years to come.

Since the turn of the millennium, vast improvements in automation, high throughput computing, and cost reduction have placed genetic sequencing technology within the reach of taxonomists searching for genetic markers to aid in their identification and classification efforts. The latest and greatest of these methods is known as next generation sequencing,[15] and a few early pieces have starting falling into place pertaining to the establishment of evolutionary pathways and affinities of our agaves. It is still early days, but the burgeoning field of sequencing analysis has the potential to completely redefine classic fundamentals of taxonomy. Stay tuned for future developments.

A. chrysantha and *A. toumeyana* share habitat in central Arizona.

CHAPTER 5

Naturally Occurring Agaves

THERE ARE TWELVE NATURALLY OCCURRING AGAVE species in Arizona. Three in particular, *A. chrysantha*, *A. palmeri*, and *A. parryi* are central players in our pre-Columbian cultigen drama, with substantial overlapping characteristics and ranges. While others are more peripheral, all represent critical aspects of the landscape in which our cultigens were developed, gardened, and farmed. And it's important to recognize that indigenous Native Americans relied heavily upon these very same naturally occurring agaves for millennia to meet a great many needs before some were polished and honed via sly horticultural technique into finely crafted anthropogenic cultigens. Besides, I've spent a lot of time out there with the Agavaceae natives, and may as well impart what few nuggets of wisdom I've managed to glean.

A. ×ajoensis from Organ Pipe Cactus National Monument.

AGAVE ×AJOENSIS W. C. Hodgson (2001) is a very rare hybrid, found only at a single locale within Organ Pipe Cactus National Monument in extreme southern Arizona, where *A. deserti* var. *simplex* and *A. schottii* share habitat at 1,000 m (3,300′) elevation. Once considered conspecific with *A. schottii* var. *treleasei*, *A. ×ajoensis* was discovered to be distinct, and subsequently described by Wendy C. Hodgson (2001).[16]

Not surprisingly, this cespitose habitat hybrid looks a great deal like *A. schottii* var. *treleasei*. Plants are 40 cm (16″) tall and wide, with light green, ascending 15–30 mm (0.6″–1.2″) wide leaves. Not only does this attractive agave lack the marginal spines of one parent, it lacks the midstripe and filiferous quality of the other as well. Small groups of 3–10 plants present along the outskirts of *A. schottii* patches where it shares habitat with *A. deserti* var. *simplex*, and 2.5–3.5 m (8′–12′) spicate to racemose stalks produce bright yellow 30–40 mm flowers in May.

Ploidy refers to the number of complete sets of chromosomes in a cell, which is key toward determining the reproductive ability of plants. Many agaves are diploid. *A. ×ajoensis* is a triploid,[17] the result of tetraploid seed donor *A. schottii* crossing with diploid pollen donor *A. deserti* var. *simplex*. That being the case, *A. ×ajoensis* morphology leans rather definitively toward *A. schottii*, which contributes twice as much genetic material to its hybrid progeny as *A. deserti* var. *simplex*. And since it is effectively sterile, asexual reproduction via rhizome is the order of the day.

One little *A. ×ajoensis* cluster seems distinct from all others. Plants are at least twice the size, and sport a few small marginal spines. The overall appearance suggests a more intermediate hybrid than normal *A. ×ajoensis*, which leans toward *A. schottii*. Bloom stalks from this odd bunch also lean more toward *simplex* than those of normal *A. ×ajoensis*. As this cluster may demonstrate, triploid sterility is not necessarily absolute, since this little

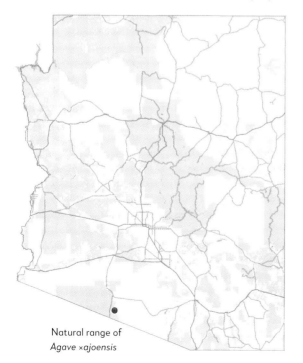

Natural range of *Agave ×ajoensis*

Naturally Occurring Agaves

A. ×*ajoensis* bunch certainly appears backcrossed with *A. deserti* var. *simplex*. In fact there are certain unlikely paths, including unreduced gamete production, by which triploids might manage to generate fertile seed with a diploid partner, but the details of such complexities are beyond the purview of this text.

Unfortunately, *A.* ×*ajoensis* is all but impossible to find in the trade, and the only known habitat population is currently under siege from eriophyid mites (aka agave mites), a microscopic pestilence that scars leaf surfaces, interferes with reproduction, and even kills its hosts. Agave mite is discussed in more detail in Chapter 10.

×*ajoensis* colonizing a small rocky outcrop.

A. × arizonica generates a bloom stalk at the end of a long, tough life.

AGAVE ×ARIZONICA Gentry & Weber (1970) was originally discovered in the 1960s, and Howard Scott Gentry himself apparently failed to realize he had come across the naturally occurring hybrid, *A. chrysantha* × *A. toumeyana* var. *bella*, instead aligning the taxon with *A. utahensis*, presumably due to a similar racemose inflorescence. *A. ×arizonica* is roughly 35 cm (14″) tall and broad with dark green 3–4 cm (1.2″–1.6″) wide leaves. Newer leaves are edged with a prominent red margin, which may be lined with small to vestigial marginal spines. Bloom stalks are 3–4 m (10′–13′) tall, and produce small (25–30 mm) yellow blooms from March through June. Like its parents, *A. ×arizonica* is considered hardy to –12°C (10°F).

A. ×arizonica has an interesting 30+ year history regarding its legal status. Originally described in 1970 as a species, subsequent field investigation suggested otherwise as it became increasingly clear that the status of *A. ×arizonica* was more properly resigned to that of hybrid rather than species. This might have been of little import, except that *A. ×arizonica* had found its way to the US Endangered Species list in 1984. Hybrid status, in and of itself, was not cause to remove this extremely rare agave from consideration as an endangered species, as many modern plant species undoubtedly arose from hybridization.

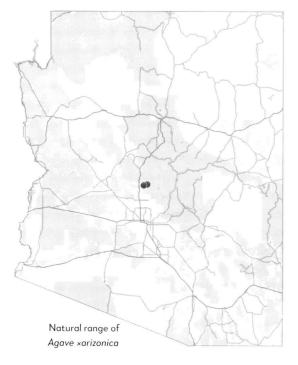

Natural range of *Agave ×arizonica*

Unfortunately, pollen viability is extremely low, and *A. ×arizonica* seems poorly adapted to the harsh environment in which its parents reside. Individual plants are found in very small numbers across such a vast area that the chance of non-clonal fertilization is beyond remote. And though decidedly cespitose under favorable conditions, inadequate hydration resulting from prolonged drought has taken an unfortunate toll on habitat populations since its discovery. In the end, it was officially delisted in 2006, once it was determined that

A. ×*arizonica* offset from a plant, which had bloomed and perished a few years prior.

the plant exists only as an impromptu hybrid, with no stable populations to protect.

Indeed, this may be the world's rarest named agave, as I know of only two living examples in habitat, miles apart in very remote sections of the New River Mountains. I had also found specimens which have since flowered and died, without leaving any progeny, and sadly bore witness the last gasp of a now regionally extinct population at Sierra Ancha.

Unusual examples of a closely related, but distinct *A.* aff ×*arizonica* ("aff" = taxonomic lexicon, short for "affinis" meaning "similar to") hybrid are found southeast of Payson. Hybrids here also feature *A. chrysantha* × *A. toumeyana* var. *bella*, but the local *A. toumeyana* population seems to have some ploidy issues. In this instance, tetraploid (rather than diploid) var. *bella* contributes twice as much genetic material as diploid *A. chrysantha* to the hybrid, resulting in distinct and sterile triploid progeny that leans more heavily toward *A. toumeyana* morphology than does normal *A.* ×*arizonica*.[18]

Though scarce in habitat, *A.* ×*arizonica* offsets freely when adequately watered in cultivation, and is occasionally available from Desert Botanical Garden plant sales, and perhaps less commonly in the retail nursery trade. It is an extremely attractive small agave, equally at home when cultivated in ground or containers.

A. ×*arizonica* blooms are an odd blend of styles from its two parents.

One-of-a-kind *A.* ×*arizonica* × *A. toumeyana* var. *bella* hybrid.

A. chrysantha from the northern end of its range.

AGAVE CHRYSANTHA Peebles (1935) is very common across central Arizona and named for its striking golden/orange blooms (Greek: chrys = gold, anth = flower). It often grows in the same general areas as *A. parryi*, with which it readily hybridizes, but in far more prodigious numbers, and often at lower elevation and more rocky terrain. Its sweet spot is 1,100–1,400 m (3,600′–4,600′), but I occasionally find it as low as 700 m (2,300′), and only rarely above 1,700 m (5,600′). At 80–180 cm (2.6′–6′), it is generally larger than *A. parryi*, and more impressively armed. Lanceolate to linear-lanceolate leaves average 8–10 cm (3″–4″) wide. *A. chrysantha* offsets now and again, but is predominantly solitary, and manages temperatures as low as –12°C (10°F).

Tall paniculate stalks emerge in spring, then typically reach 5–8 m (16′–26′) and flower from late June though mid to late July. Flowers are on the small side (35–50 mm), and feature stiff, erect tepals, which help retain nectar, to the delight of pollinating hummingbirds.

A. chrysantha is a wildly diverse species, heavily introgressed with *A. parryi* toward the north, with wide, glaucous blue leaves, and decidedly more *palmeri*-like toward the opposite end of its range, with green leaves lengthening and narrowing as we head south. In fact, toward the southern end of its range, *A. chrysantha* is so strikingly similar to *A. palmeri* that it was formally described as *A. palmeri* var. *chrysantha* at one time, an errant name that has dogged it for years. Despite similar vegetative characteristics and flower morphology, recent next generation sequencing analysis conducted by Andrew Salywon and Wendy Hodgson at Desert Botanical Garden actually suggests a startling lack of affinity between *A. chrysantha* and *A. palmeri*, deferring instead to convergent evolution as the backbone of this dynamic.

Left to its own devices, and free from *A. parryi*, *A. palmeri*, and pre-Columbian cultigen

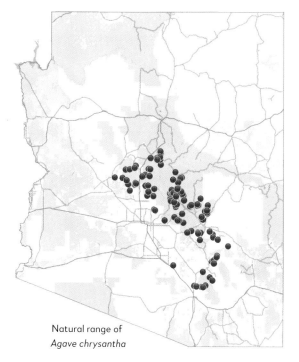

Natural range of *Agave chrysantha*

Black-chinned hummingbirds (*Archilochus alexandri*) are favored *A. chrysantha* pollinators.

influence in a remote section of central Arizona, *A. chrysantha* grows to absolutely leviathan proportions. Many of these stunning agaves grow to 250 cm (8′) across and sport bloom stalks in the 9–10 m (30′–33′) range. These may be the second largest native habitat agaves in the US, trailing only the poorly known Florida taxon, *A. neglecta*.

A. chrysantha is frequently found in the vicinity of pre-Columbian cultigens throughout central Arizona, and is easily confused with *A. delamateri* and *A. verdensis,* in particular. The interactions of *A. chrysantha* with our pre-Columbian cultigens are extensive, and examined in substantial detail in Chapter 8. Salado and Hohokam would have encountered *A. chrysantha* regularly as the omnipresent agave standard bearer throughout much of the region.

A. chrysantha hybridizes with *A. parryi, A. toumeyana, A. schottii, A. murpheyi, A. delamateri*, and *A. verdensis. A. chrysantha* × *A. parryi* hybrid populations absolutely dominate the landscape in several areas across central Arizona, typically occupying habitat intermediate between the sweet spots of each. One intriguing locale features evidence suggestive of *A. mckelveyana* introgression, but no obvious F1 hybrids, since the current range of each approaches no closer than 5 km (3 miles).

A. chrysantha, from the land of giants in central Arizona. This pair measures 4 m across.

Hundreds of A. chrysantha in s

Extremely rare striated A. chrysantha.

Large, banded A. *deserti* var. *simplex* from the Sierra Estrella Mountains.

Naturally Occurring Agaves

The range of *AGAVE DESERTI* VAR. *SIMPLEX* Gentry (1978) is quite apart from that of our pre-Columbian cultigens, preferring especially hot, dry terrain as low as 300 m (1,000') others could not manage. But its range does intersect with that of the Hohokam, and they may have demonstrated some substantial interest. *A. deserti* var. *simplex* is considered one of the more edible wild agaves, and is still harvested and consumed today by members of the Tohono O'odham Nation, some of whom consider themselves descendants of the Hohokam. Widespread utility is also suspected of a broad range of ancient Native American groups that were less reliant upon agriculture.

A. deserti var. *simplex* is a small to middling, mostly solitary, (usually) blue agave, 50–100 cm (20"–40") across, with rigid lanceolate leaves 6–10 cm (2.3"–4") wide. Medium-sized (40–60 mm) flowers grace 4–6 m (13'–20') panicles mid-May through mid-June. We find *simplex* scattered across western Arizona from the Mexican border all the way to some 50 km (31 miles) north of Alamo Lake in Arizona, and into eastern California. Individual plants are more widely scattered and difficult to find than many other regional agaves. Ocotillo (*Fouquieria splendens*) is a favorite nurse plant. *A. deserti* var. *simplex* can manage elevations as high as 1,400 m (4,600'), and is hardy to −15°C (5°F).

I have explored the haunts of *A. deserti* var. *simplex* extensively in my travels, and discovered an interesting phenomenon while wandering about the type locality at the Harquahala Mountains. Specimens with distinct vegetative characteristics commonly grow side-by-side. Some plants resemble *A. deserti* from Southern California and Baja, while others bear a striking resemblance to *A. parryi*, though marginal spines are not always as subdued. Some of these agaves feature the yellow flower buds of *A. deserti*,

Natural range of *Agave deserti* var. *simplex*

from Organ Pipe Cactus National Monument.

while others are the same bright crimson as those of *A. parryi*. Many are intermediate.

    Further investigation has yielded a loose but discernable pattern. Low-lying, harsh desert populations typically demonstrate characteristics more closely aligned with those of *A. deserti*, while populations along higher elevations lean more toward *A. parryi*, which makes sense if *A. parryi* or an *A. parryi*-like forebear is in the ancestral genome of *simplex*. This unusual dichotomy extends toward both vegetative and inflorescence characteristics, but flower morphology is rather more consistent. Perhaps the unusually even distribution of these characteristics across the unique agave population at Harquahala inspired H. S. Gentry to select it as the type locality. *A. parryi*-like characteristics are

Beautiful *A. parryi*-like examples of *A. deserti* var. *simplex*.

so extreme in one peripheral population, distinct taxonomic recognition may be warranted.

Andrew Salywon of the Desert Botanical Garden reports that *A. deserti* var. *simplex* may soon assume status as a distinct species, apparently not so closely related to *A. deserti* as once believed.

"*Agave simplex*," as it will soon be known, hybridizes with *A. schottii*.

A large and isolated agave population appears intermediate between *A. deserti* var. *simplex* and *A. parryi*.

A. deserti var. *simplex* in all its glory, from the type locality at Harquahala Mountain.

A. mckelveyana catching a few rays in western Arizona.

AGAVE MCKELVEYANA Gentry (1970) is closely related to *A. deserti* var. *simplex*, often preferring the markedly distinct habitat of pinyon-juniper woodland from 900–1,400 m (2,900'–4,600'). It is named in honor of Susan Delano McKelvey, the twentieth century botanist, horticulturist, and cousin to president Franklin Delano Roosevelt. Mrs. McKelvey, as it turned out, played a pivotal role in the initial discovery and history of the pre-Columbian cultigen *A. delamateri* and definitely left her mark on the Arizona botanic landscape.

Like *simplex*, it is found primarily in western Arizona, but farther north and usually at higher altitude. This is the smallest of the paniculate agaves found in the region, averaging perhaps 40–50 cm (16"–20") tall and wide, with 3–5 cm (1.2"–2") wide linear to lanceolate leaves. Bright yellow flowers appear along 3–4 m (10'–13') panicles through May and bear a strong resemblance to those of *A. deserti* var. *simplex*, but are considerably smaller at 30–40 mm. This is one of our hardiest agaves, able to withstand temperatures as extreme as −20°C (−5°F).

It seems entirely likely that some manner of common ancestry manifests between *A. mckelveyana* and *A. utahensis* var. *utahensis*. Compared to *A. deserti* var. *simplex*, *A. mckelveyana* is smaller, greener, and more cespitose, qualities well aligned with *A. utahensis*. And it appears that some small percentage of *A. mckelveyana*, no more than 3%–4%, sport the same distinctive deltoid marginal spines as *A. utahensis*. And compelling evidence suggests that *A. utahensis* is actually an ancient *Agave* × *Littaea* hybrid.

In a remote section of west-central Arizona *A. mckelveyana* and *A. utahensis* var. *utahensis* share habitat, and a few obvious hybrids grow directly alongside *A. mckelveyana*. These moderately cespitose hybrids are on the small side, and produce 2–3 m (6.5'–10') racemose stalks. A few kilometers away, occasional racemose stalks as tall as 8 m (27') present on unusual 60–90 cm (2'–3') agaves along

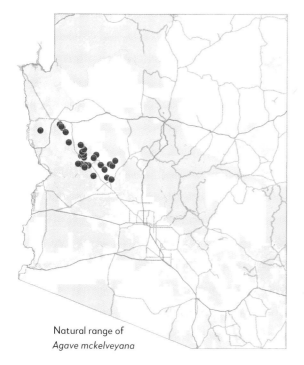

Natural range of *Agave mckelveyana*

A. mckelveyana finishing up its bloom cycle

A. mckelveyana × A. utahensis hybrid strutting its stuff.

hillsides otherwise populated exclusively by *A. mckelveyana*. Over the course of investigating this phenomenon, a few small relict *A. utahensis* patches were found scattered across the region, and though none are close enough to suggest F1 introgression, this suggests that *A. utahensis* once enjoyed a far broader distribution in the area than it does today.

A. mckelveyana occasionally shares habitat with our westernmost *A. parryi* populations, with which it hybridizes. In these areas, *A. parryi* is often seen stretching out for sun in open terrain, while *A. mckelveyana* is hiding under bushes, with only bloom stalks to betray its presence. In other areas, *A. mckelveyana* shares habitat with the pre-Columbian cultigen *A. phillipsiana*. These species grow in close proximity along the very same hillsides south of Prescott, but there is no apparent interaction, since bloom times are three months apart.

A. mckelveyana readily hybridizes with both *A. parryi* and *A. utahensis*.

A. palmeri from the Dragoon Mountains in southeastern Arizona.

AGAVE PALMERI Engelmann (1875) starts up where *A. chrysantha* stops, widely distributed across the southeast quadrant of Arizona and extending well into northern Sonora, Mexico. A good-sized agave, Arizona's largest on average, *A. palmeri* features long, 8–10 cm (3″–4″) wide glaucous green lanceolate leaves, and typically grows to 180 cm (6′), but occasionally much larger. In addition to distinct bloom color and habitat, small, numerous marginal spines serve to distinguish these plants from more heavily spined *A. chrysantha* populations. *A. palmeri* prefers grassy terrain along gentle slopes and flats from 1,200–1,800 m (4,000′–6,000′), and can manage temperatures down to about −10°C (15°F).

Blooms along 4–9 m (13′–30′) paniculate stalks open late June through mid-August. Flowers are a little larger than those of *A. chrysantha* and feature similar erect tepals, though *A. palmeri* tepals sport distinct callouses, dried brown or reddish spots on their tips, which is an unusual feature, unique among naturally occurring agaves in the area and a signature characteristic of four of our five pre-Columbian cultigens. And a real oddity, *A. palmeri* may be the only known agave species with more than one color scheme for blooms, ranging from soft yellow to green/cream with bright red filaments and anthers. These distinctly colored variants occasionally stand atop stalks side-by-side, just a couple meters apart. Each color scheme aligns with one or more of our pre-Columbian cultigens, with the more common green and red color scheme quite similar to that of the cultigen *A. delamateri*.

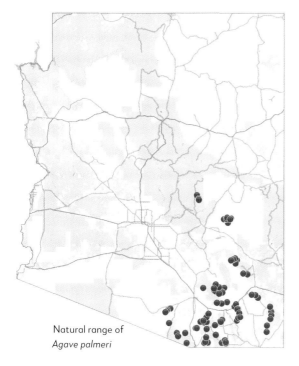

Natural range of *Agave palmeri*

The flowers of *A. palmeri* are specifically oriented to attract nectar feeding bats as favored pollinators. Nocturnal nectar and pollen production also favor bat pollinators. The symbiotic relationship between *A. palmeri* and migratory long-nosed bats has been studied extensively, and a long-term trend of poor *A. palmeri* seed set had been linked to declining bat populations, primarily of the genus *Leptonycteris*.[19] This decline

A. palmeri in favored habitat.

was largely associated with roost site disturbance, agave habitat loss, and *A. tequilana* cultivation practices south of the border, and the lesser long-nose bat, *Leptonycteris yerbabuenae*, was added to the US Endangered Species list in 1988. In May of 2018, just thirty years later, the endangered species was triumphantly delisted, as populations rebounded dramatically from a low of perhaps 1,000 individuals to an estimated 200,000 bats living in 75 roosts across the southwestern US and Mexico. This is the very first bat species ever removed from the US Endangered Species list, so a real triumph for all concerned, including bats, humans, and a host of prominent desert plants that rely on these important pollinators.

Although vegetative characteristics of some of our cultigens align well enough with *A. chrysantha* to occasionally confound identification attempts from a discreet distance, DNA analysis from Andrew Salywon of the DBG, and flower morphology suggest greater affinity with *A. palmeri*. Which makes sense, since *A. palmeri* has a long history of utility as a food and fiber source by indigenous Native Americans. Its expansive range and broad use suggests an ideal candidate to select and hybridize toward an end of further refinement.

Perhaps even more so than *A. chrysantha*, *A. palmeri* is stupendously diverse, with populations of miniatures perhaps 60–80 cm (23"–32") across, not far from the Mexican border, and other populations teetering so far on the edge that they may merit distinct taxonomic recognition.

A single pre-Columbian cultigen skirmish manifests as a subtle sideswipe along the far eastern edge of *A. delamateri* range in Salado territory. This isolated *A. palmeri* population is at the very northern edge of its range, some 130 km (80 miles) distant from its nearest conspecific neighbors to the south. Whether a relict population or an anthropogenic distribution gone feral I cannot say with certainty, though I rather favor the former premise. For now, the origins of this population remain uncertain.

A. palmeri is known to hybridize with *A. parryi* and rarely with *A. parviflora*. A few *A. palmeri* drift down to swap genes with *A. chrysantha* in the valley separating the Rincon and Santa Catalina Mountains, just east of Tucson.

Most *A. palmeri* blooms feature bright red filaments.

Distinct A. palmeri bloom color schemes displaying side-by-side.

A. parryi var. *parryi* growing at high elevation in the Santa Catalina Mountains east of Tucson.

Naturally Occurring Agaves

AGAVE PARRYI Engelmann (1875) is Arizona's one and only montane agave, usually found from 1,300–2,400 m (4,200'–7,900'), often in sky island pinyon-juniper woodland, but also in grassy flats, across a wide swath of Arizona from its southeast corner all the way north to old Route 66 near Peach Springs, some 450 km (280 miles) away. Range also extends south into Mexico, and evidence suggests a far greater US distribution at some point in our distant past. That evidence presents in the form of apparent *A. parryi* introgression in other agave species, including *A. chrysantha* and *A. mckelveyana*, in areas in which *A. parryi* is now entirely absent.

A. parryi usually glows bright blue, but sometimes green, and reaches up to 120 cm (4') across, though many are only half that size. Its general appearance inspires the common name, artichoke agave. Moderately cespitose and hardy to at least −18°C (0°F), these agaves are typically compact and stout, with very stiff, rigid leaves and formidable terminal spines.

There are five recognized *A. parryi* taxa, three of which reside in Arizona. These include var. *parryi*, var. *huachucensis,* and var. *couesii*. *A. parryi* var. *couesii* is the northernmost variety and features the narrowest leaves. Leaves generally widen as we head south through var. *parryi* and var. *huachucensis* territory. Two additional taxa, var. *truncata* and subsp. *neomexicana*, are found respectively near the Durango-Zacatecas border in Mexico, and New Mexico.

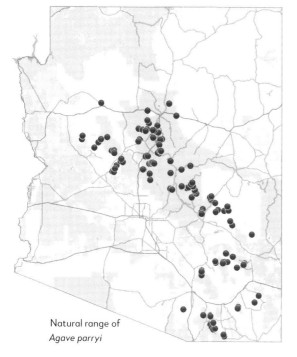

Natural range of *Agave parryi*

There is a great deal of vegetative characteristic overlap in these taxa, and some seem rather more distinct than others. *A. parryi* subsp. *neomexicana*, for example, is easily confused with var. *couesii* or var. *parryi* until it blooms, when a notably less robust inflorescence emerges. Floral characteristics are also marginally distinct, enough so that Gentry considered subsp. *neomexicana* a separate species altogether. Prior to 1939, var. *couesii* was also considered a separate species (*A. couesii*).

Unusually large 1.4 m *A. parryi* var. *parryi*, happy as can be in the New River Mountains.

Long, narrow leaves are characteristic of *A. parryi* var. *couesii*.

Other than the distinctively long, narrow leaves of many var. *couesii* specimens, variants of size, color, and leaf shape are found among all Arizona populations, rendering variety identification nearly impossible without locality information.

Bloom stalks are shorter than those of many other regional agaves at 4–6 m (13'–20'), but extremely thick and sturdy, often featuring tightly bunched panicles that present a distinct appearance. Bright yellow flowers open from early June through early July, and present a stark and breathtaking contrast to glowing crimson buds.

A. parryi demonstrates a moderately distinct bloom strategy than other regional agaves, as an apparent

Naturally Occurring Agaves

adaptation to montane existence. Most agaves start blooming at the bottom of the inflorescence, with open flowers slowly working their way skyward over a period of weeks. In contrast, at least 90% of *A. parryi* blooms open over a day or two, leaving only a small fraction at the top to continue at a more typical pace. One negative consequence is that this wild bounty of riches might overwhelm available pollinators, thus reducing germination rates. This however is apparently offset by a promise of more rapid development and fruition. Since *A. parryi* prefers higher elevations than other regional species, it is more exposed and vulnerable to fierce summer monsoon storms, which can topple and uproot developing stalks and seed. In this case, faster is apparently better, as it is Gaia herself who favors more successful strategies via natural selection. Another montane species, *A. parrasana*, from Coahuila, Mexico, also has an extremely thick, sturdy stalk and employs a similar bloom strategy, in apparent deference to high altitude concerns.

A. parryi was widely cultivated by Hohokam, Sinagua, and Salado as a valuable source of fiber and food, and is commonly found in the company of the pre-Columbian cultigen, *A. phillipsiana*, and less commonly with *A. delamateri*. This is addressed in substantially more detail in Chapter 9.

A. parryi is known to hybridize with *A. chrysantha*, *A. mckelveyana*, *A. palmeri*, and *A. schottii*.

A. parryi blooms are among the most striking of all agaves.

A. parviflora never disappoints.

A. parviflora, just a stone's throw from the Mexican border.

AGAVE PARVIFLORA Torrey (1859) is a small but extremely attractive and desirable agave that bears a fair resemblance to *A. toumeyana*. It grows along both sides of the US-Mexican border at 1,200–1,500 m (3,900′–4,900′), and is the only agave currently found on the US Endangered Species list, considered threatened. Dark green oblong linear leaves are filiferous and boldly marked with striking white imprints. Spicate stalks are a little shorter than those of *A. toumeyana*, and blooms, though smaller, are quite similar and emerge in June. Hardiness is to –15°C (5°F).

The species is named for its exceedingly small flowers (Latin: parvi = small, flora = flower), measuring 13–15 mm. I have found about twenty habitat sites, none with more than 200–300 individuals, and many with less than ten, each plant perhaps 18 cm (7″) across or less.

As similar as *A. parviflora* and *A. toumeyana* var. *bella* seem in culture, they are startlingly distinct in habitat. *A. toumeyana* is a decided extrovert, growing in thick patches on open ground or rocky terrain screaming "Look at me!" while the less cespitose *A. parviflora* is more the shy girl sitting in a corner, desperately hoping no one asks her to dance. Of course we do, because she is exquisitely lovely.

One thing about this species that always strikes a note is its unusual post-mortem persistence. Most agaves deteriorate pretty quickly after blooming, but I regularly see very attractive *A. parviflora* specimens that had bloomed at least two years prior. Unfortunately, new offsets are apparently not produced during this time. Another standout feature, as previously alluded to, is the startling scarcity of these plants, even at places in which they occur. Many of the sites I've documented contain only a handful of plants, and even the busiest sites feature extremely sparse populations when compared to other agaves in the region.

A. parviflora rarely hybridizes with *A. palmeri*, with which it shares habitat.

Natural range of *Agave parviflora*

A. schottii blanketing a rocky outcrop.

AGAVE SCHOTTII Engelmann (1875) may actually be the world's least attractive agave, and its common name of "shin dagger" is well earned as stiff, sharp filiferous leaves are not always easily discerned while hiking through grassy fields. They do have a saving grace however, putting on quite the show in June with beautiful bright yellow blooms gracing 2–3 m (6.5'–10') spicate stalks.

When growing in favored habitat, this cespitose agave forms a dense clump of up to 50 cm (20"), and can blanket gentle grassy slopes in enormous numbers. Leaves are stiff and narrowly linear, less than 1 cm wide on average, far too narrow to serve as adequate canvas for white leaf impressions, which are not always readily apparent unless carefully examined. *A. schottii* sometimes plays habitat hopscotch with the likes of *A. palmeri*, and even *A. parryi* across southeastern Arizona. *A. schottii* is encountered between 750–1,900 m (2,400'–6,200'), and hardiness is to at least –12°C (10°F).

This diminutive agave is closely related to *A. toumeyana* and *A. parviflora*,[20] and was undoubtedly encountered with some regularity by Hohokam and Salado while foraging through hilly terrain. *A. schottii* is very high in sapogenins, and has long been utilized as amole, a reference to agaves, yuccas, and other plants in which roots or other plant parts are used as detergent. Sapogenins are steroidal compounds derived from saponins, and protect plants against a variety of pathogens and herbivores. A high sapogenin content renders *A. schottii* unpalatable to wild grazers and livestock.

A report of oven-baked *cabeza* consumption by the indigenous Tarahumara people of northwestern Mexico suggests the possibility of similar use by early Native Americans.[21] Pit roasting amole may seem unlikely given its high sapogenin content, but the heating process goes some substantial distance toward breaking down these bitter-tasting compounds, and the small *cabezas* can be roasted to good effect within

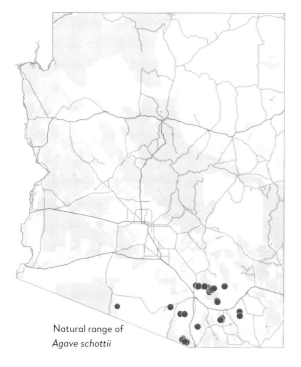

Natural range of *Agave schottii*

A. schottii growing along a rocky perch.

a single evening. So perhaps not as healthy, but greater speed and convenience. Sound familiar?

A. schottii hybridizes with *A. chrysantha*, *A. deserti* var. *simplex*, and *A. parryi*, and all of these rare hybrids are really quite alluring in their own right. Persistent rumors of *A. palmeri* introgression remain unconfirmed.

A. schottii plants may not catch your eye, but bloom stalks will!

A. schottii var. *treleasei* is easy to spot, but extremely scarce in habitat

Naturally Occurring Agaves

AGAVE SCHOTTII* VAR. *TRELEASEI Kearney & Peebles (1939), is currently regarded as a variety of *A. schottii*. I include it here as a distinct entry because it is actually an *A. schottii* × *A. chrysantha* hybrid[22] and should be formally named anew as the hybrid taxon *A.* ×*treleasei*. It is nearly as scarce as *A.* ×*arizonica*, growing in a few small clumps at 1,100–1,900 m (3,600′–6,200′) in the Santa Catalina Mountains east of Tucson. This small cespitose agave may have been somewhat more common at one time, as an account published in 1938 clearly describes it as "used by various Indian groups as amole."[23] The few examples that survive today can hardly be found, let alone used. Despite that reference, sapogenin content seems less pronounced than in *A. schottii*, as some of the few examples encountered have been grazed, presumably by deer.

This 35–55 cm (13″–22″) dark green agave actually looks rather delicate and vulnerable, with 15–30 mm (0.6″–1.2″) wide leaves and no protective marginal spines. Like *A.* ×*ajoensis*, with which it was once considered conspecific, *A. schottii* var. *treleasei* is a sterile triploid. Unlike its seed donor parent, which it most resembles, there is no midstripe and leaves are not filiferous. When cultivated, *A. schottii* var. *treleasei* offsets prolifically, and is available in the retail nursery circuit from time to time. It should be equally hardy as its parents, good to –10°C (15°F).

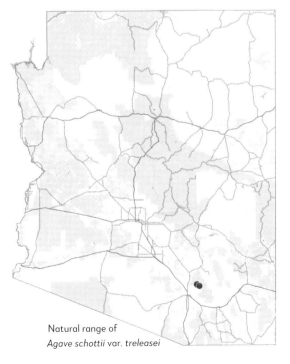

Natural range of *Agave schottii* var. *treleasei*

A. schottii var. *treleasei*, catching some morning sun

A. toumeyana from central Arizona.

Naturally Occurring Agaves

AGAVE TOUMEYANA Trelease (1920) occupies roughly the same terrain as *A. chrysantha* across central Arizona, but prefers a different substrate, so are not often seen growing side-by-side. More often they play habitat hopscotch, with *A. toumeyana* frequently less common and certainly more selective about where it sets down roots. Far more concerned with substrate than elevation, *A. toumeyana* features a broad range of 900–1,800 m (3,000′–6,000′).

This very attractive *Littaea* agave produces a 2–3 m (6.5′–10′) spicate inflorescence, with green and cream flowers that mostly open through June. Bold white markings on dark linear lanceolate leaves are especially pleasing to the eye. Leave edges are adorned with all manner of white thread, and average 1–2 cm (0.4″–0.8″) wide. This sly little agave has an unusual trick up its sleeve. A small number, perhaps 5%–10%, forego the late spring ritual and bloom in late autumn instead. It is always a little startling to encounter agave blooms in mid to late November in Arizona. Flowers are designed primarily for insect pollination, and are oft frequented by apian visitors, including bumblebees and carpenter bees.

These cespitose agaves are hardy to –12°C (10°F), and some are very small at 15 cm (6″) across but form large colonies of offsets, not infrequently forming fairy rings. Imagine a single small cespitose agave surrounding itself with offsets. A few years pass, and the original agave blooms and perishes. Offsets continue to develop while producing their own offsets, radiating away from the original agave husk, which is still occupying space. A few more years pass and they, too, bloom and perish. On and on until decades later, we're looking at a 3 m (10′) donut, comprised of dozens or even hundreds of small agaves.

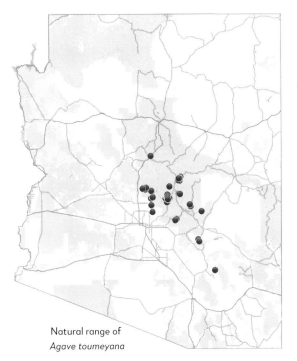

Natural range of *Agave toumeyana*

There are two recognized taxa. *A. toumeyana* var. *toumeyana* is a larger plant reaching 60 cm (2′) with perhaps 40–60 leaves. *A. toumeyana* var. *bella* is less common and much smaller, with as many as 200 leaves. *A. toumeyana* is unusual amongst regional agaves in that

Large A. toumeyana fairy

A. toumeyana var. bella can definitely catch your eye.

A. toumeyana var. *bella* keeping company with *Dudleya saxosa*.

it appears to demonstrate variable ploidy.[24] Many/most populations are diploid, but a large community across the northern part of its range is at least partially tetraploid. Research in the area is sorely needed for further clarification.

Only three populations of *A. toumeyana* var. *bella* are generally known, one at New River, one at Sierra Ancha, and a third in the northern Mazatzal Mountains, south of Payson. That said, var. *bella* is hardly rare, as each population is rather robust. Carefully feel the base of leaf edges to discern which you have, as var. *bella* sports small marginal nubs while var. *toumeyana* is smooth all the way to the base.

A. toumeyana grows throughout ancient Salado territory, and may have met use as amole given its high sapogenin content, similar to that of *A. schottii*. *A. toumeyana* is occasionally found in the vicinity of three of our cultigens.

A. toumeyana hybridizes with *A. chrysantha*, though these hybrids are extremely rare.

Spicate inflorescence of *A. toumeyana*.

A. utahensis subsp. *kaibabensis*, more than a meter across.

AGAVE UTAHENSIS Engelmann (1871) is North America's most northerly agave, with a range extending north of the Grand Canyon and into Nevada and Utah, as its name suggests, and is tolerant of biting temperatures as extreme as −20°C (−5°F). There are four (arguably) recognized taxa,[25] two of which, subsp. *utahensis* var. *utahensis* and subsp. *kaibabensis*, reside in Arizona, with most Arizona plants found in the general vicinity of the Grand Canyon. Two additional taxa, subsp. *utahensis* var. *eborispina* and subsp. *utahensis* var. *nevadensis*, are found in California and Nevada.

A. utahensis subsp. *kaibabensis* Gentry (1982) is bound to limestone substrate and dramatically distinct from other *A. utahensis* forms. Not only is it solitary, *kaibabensis* is much larger, topping out at more than a meter across with a correspondingly impressive inflorescence of up to 5 m (16′). In the eastern part of the Grand Canyon area, where it resides, *kaibabensis* is the nearest agave neighbor of the pre-Columbian cultigen, *A. phillipsiana*. Like our cultigens, *kaibabensis* has a long history of utility by indigenous Native Americans as a source of food and fiber. I suspect subsp. *kaibabensis* as precedent over other forms of this species, with others possibly evolving via adaptive radiation and introgression with other agave species.

A. utahensis var. *nevadensis* is the smallest of this bunch. Usually bright blue and wildly cespitose, *nevadensis* can sweep across suitable habitat in astonishing numbers. In this case, suitable habitat means pretty much any manner of limestone substrate. *A. utahensis* var. *eborispina* (Latin: ebur = ivory, spina = spine) essentially presents as an offshoot of *nevadensis*, though greener, a bit larger, and far less cespitose. Terminal spines on both are oversized and especially eye-catching. Spines of var. *eborispina*, in particular, are oft exaggerated to cartoonlike proportions.

A. utahensis var. *utahensis* is the only form not bound to

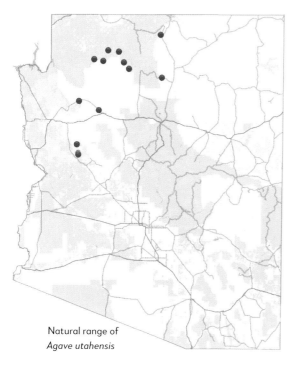

Natural range of *Agave utahensis*

A. utahensis offsets furiously to form large colonies.

A. utahensis from the very southern edge of its range.

Tall spicate inflorescence of *A. utahensis* subsp. *kaibabensis*.

A dizzying mass of long white terminal spines is an unmistakable *A. utahensis* var. *eborispina* characteristic.

limestone. All forms of *utahensis* are extremely variable, and the size of var. *utahensis* is no exception, ranging from 15–70 cm (6"–28") across. Like its close cousin, var. *nevadensis*, this form is highly cespitose, sometimes forming impressive fairy rings as large as 10 m (32') across. *A. utahensis* var. *utahensis* occasionally shares habitat with both *A. mckelveyana* and *A. parryi*, but has not been found in the vicinity of our cultigens.

All forms of *A. utahensis* feature blunt deltoid, light gray marginal spines along thick, rigid, linear lanceolate leaves. These marginal spines are utterly distinct and once seen, easily recognized in other populations.

Though considered a member of subgenus *Littaea*, *A. utahensis* demonstrates extreme variability of its inflorescence, ranging from spicate to racemose to (rarely) paniculate. Seed pod size and shape also vary dramatically. Of this, Gentry notes "with study of the natural populations, such variation becomes a species character, bringing the whole into perspective."[26] Such variability is strongly suggestive of ancient *Agave* × *Littaea* hybrid ancestry, but further study is warranted. Bloom time is typically March through May, and flowers are bright yellow.

A. utahensis var. *utahensis* is known to hybridize with *A. mckelveyana*.

A. utahensis in flower.

Hohokam ruin, perhaps 100 m above a small *A. murpheyi* stand.

CHAPTER 6

Pre-Columbian Agave Cultigens

IN SOME INSTANCES, IT SEEMS NATIVE AMERICANS found room for improvement in naturally occurring agaves. By employing techniques as familiar as selection and (presumably) hybridization, and possibly other techniques less familiar, ancient skilled horticulturalists forged a symbiotic relationship with agaves by developing and nurturing new cultigens to better suit their varied purpose. And this is where our Arizona pre-Columbian cultigens step in. In this case, the primary qualities selected and bred for appear to be fiber and food.

Most of our Arizona cultigens were likely planted some 700–1,100 years ago,[27] and share certain qualities. And it is many of these very qualities that strongly suggest an anthropogenic rather than natural origin. All are mid-sized; reproductively compromised; produce large, healthy *cabezas*; flower synchronously from a paniculate inflorescence; and feature soft, pliable leaves that are easily cut. Careful examination of domesticated agave sites occasionally yield flattened palm-sized rocks, which had been worked and sharpened for use as agave knives. It is startling how easily these knives slice through the leaves of domesticated cultigens when compared to those of naturally occurring agaves in the region. Also, our Arizona cultigens taste sweeter and four are less fibrous than naturally occurring agaves.[28] Finally, all reproduce prodigiously by vegetative means. Were that not the case, they would have long since vanished from the landscape and not be here for us to admire.

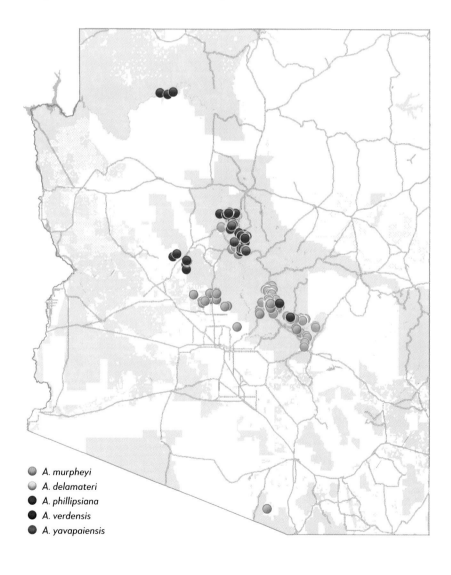

- A. murpheyi
- A. delamateri
- A. phillipsiana
- A. verdensis
- A. yavapaiensis

Sinagua pictographs from the heart of domesticated agave territory.

Pre-Columbian cultigens might escape notice were it not for their reproductive handicaps. Fully fertile cultigens would be more difficult to recognize as distinct, and might have gone feral, fomenting toward an indecipherable nucleotide stew with naturally occurring agave populations. I sometimes wonder if this scenario may have actually played out in several areas across our state.

Another shared characteristic is strikingly homogenous morphology. Of *A. murpheyi*, Gentry notes, "All plants I have seen show relatively little variation and, once the species is observed, it is readily recognized again." This lack of diversity is a signature characteristic of all our cultigens, and a prime recognition key out in the field. Experience has taught me to expect a certain degree of diversity in wild agave populations. When I encounter an isolated group of identical clones, my cultigen suspicion meter often starts ticking.

A comprehensive study completed in 2007 suggests that the pre-Columbian cultigens, *A. delamateri* and *A. murpheyi*, actually maintain a greater genetic diversity deficit than most widely distributed agricultural crops used today, and certainly far less than naturally occurring agave species. These deficit patterns align well with what is generally seen in traditional farming systems replete with money crop cultigens.[29]

Ignoring *A. murpheyi* for just a moment, others sport *A. palmeri*-like bloom characteristics in areas in which *A. palmeri* is either largely or completely absent. Three of the five grow in *A. chrysantha* territory, while *A. phillipsiana* is frequently found growing alongside *A. parryi*. Cultigens, which sport *A. palmeri*-like bloom characteristics have proven far more challenging to identify within the range of *A. palmeri*, for obvious reasons. As yet, none are described.

In any case, let's not get too far ahead of ourselves, and introduce our five formally described pre-Columbian cultigens.

A. murpheyi growing a short distance from an ancient Hohokam habitation site.

Pre-Columbian Agave Cultigens

AGAVE MURPHEYI (Hohokam agave) was originally described in 1935 by Frederick Gibson,[30] and named for William Calvert Murphey of Superior, Arizona, who as an amateur observer of native plants and animals first discovered the species. He noticed that it was unlike other agave species in the region due to its unusual bulbiferous inflorescence and leaf shape.

A. murpheyi is the very first Arizona agave subsequently recognized as a pre-Columbian cultigen. This came to light as botanists and archaeologists collaborated in the mid-1980s to uncover an association with ancient Hohokam archaeological sites.[31] Of *A. murpheyi*, Gentry writes, "Some of the clones appear to have been associated with old Indian living sites,"[32] so he may have suspected its status as domesticated, but apparently wasn't ready to commit, probably for lack of evidence at the time he penned his great tome.

Outside the valley, *A. murpheyi* is rare and nearly unknown, but in the Phoenix area groups of these 80–110 cm (31"–43") plants with ascending, linear acuminate, dark green leaves are a common sight, often gracing medians and parkways. Small regular marginal spines are distinctively short and stout. The fate of bloom stalks, which emerge mid to late autumn, is somewhat more reliant upon weather conditions than other regional agaves. In mild winters,

One of only a handful of remaining naturally occuring populations of *A. murpheyi*.

Developing A. murpheyi bulbils.

Close-up of A. murpheyi flowers.

stalks may continue to develop without pause, but most years the first cold snap halts growth in its tracks, resuming without harm later in the season after a few dormant weeks. Unusually harsh winters can damage and even abort the flowering process. Undamaged stalks eventually attain a height of 3–4 m (10′–13′), with small 30–40 mm green flowers with yellow filaments and anthers opening early to mid-spring.

The wide proliferation and use of *A. murpheyi* across the Phoenix area is a direct result of its close proximity and bulbiferous nature. Plantlets form along the inflorescence where most agaves would sport developing seed pods. These bulbils continue to grow while attached to the stalk for many months, eventually rendering the stalk top-heavy and unwieldy. Ideally, the right summer monsoon storm comes rolling in and topples the stalk. Any bulbils fortunate enough to make contact with soil can grow roots and establish, as bulbils are extremely robust, aggressive growers for at least their first few weeks.

Though rare and at high risk of in situ extinction, *A. murpheyi* is still hanging on at Sierra Ancha, the Tonto Creek area, and close to the Mexican border where it was reportedly once prevalent, but most remaining sites are now found at 500–900 m (1,600′–3,000′) around

A. murpheyi in bloom

Lake Pleasant and the New River area in startlingly harsh habitat that would turn many agaves to ash. *A. murpheyi* is not found at higher elevations, since it is hardy to only −7°C (20°F). It is the only Arizona pre-Columbian cultigen not endemic to our state, having also been found at several sites in northern Mexico.

A. murpheyi may be the oldest of our five described cultigens, and is certainly the one we know the most about. Which is not to say it is unquestionably the oldest pre-Columbian cultigen to set roots in Arizona soil. It may have had predecessors that no longer survive—there is no way of knowing. We do know, or at least find convincing evidence that *A. murpheyi* was developed in Mexico, and subsequently brought to Arizona, possibly through trade.

Not sterile, but reproductively compromised, *A. murpheyi* can produce small quantities of seed along with copious numbers of bulbils, small plantlets that develop directly on bloom stalks. Similar appearance and bulbil production have long led to speculation that *A. murpheyi* may be an *A. angustifolia* hybrid, which reports of recent next generation sequencing analysis conducted by Andrew Salywon and Wendy Hodgson at the Desert Botanical Garden apparently confirm. From there it gets a bit sketchy, but *A. angustifolia* was probably hybridized with other agaves that render *A. murpheyi* less vulnerable to subfreezing temperatures.

Large numbers of clam and seashell artifacts from the Gulf of California have been discovered at archaeological digs across Hohokam territory.[33] Shells were often decorated and used as beads, bracelets, pendants, and other forms of jewelry, suggesting regular trade with indigenous pre-Columbian cultures once occupying lands along Mexico's west coast, the very place *A. angustifolia* calls home. And *A. angustifolia* is also tied to other putative hybrids or cultigens, including *A. fourcroydes* and *A. tequilana*.[34]

As we examine pre-Columbian cultigens, it is important to consider the reasons they were cultivated. We know indigenous Native Americans used agave as a source of food, fiber, drink, and building material, but what advantages are bestowed by these specific cultigens over naturally occurring agaves? One obvious advantage of *A. murpheyi* is bloom time. Most naturally occurring Arizona agaves have a fairly narrow bloom period, starting from mid-May through mid-August, dependent upon location and species. But *A. murpheyi* blooms in early spring, so it provides a nutrient source at a time native agaves do not.[35] That is one big advantage, and bulbil production is another. What could be easier than restocking gardens and farms with hundreds of well-developed, weather-toughened little plantlets? Much easier than seed, and greater convenience and quantity than offsets.

Pre-Columbian Agave Cultigens

AGAVE DELAMATERI (Tonto Basin agave) was officially added as our second Arizona pre-Columbian cultigen in 1995. Originally discovered by Susan D. McKelvey, and named *A. repanda* by famed botanist William Trelease in 1929, he declined to publish for reasons unknown, and the ghosted taxon was subsequently misidentified by Gentry, who at one time considered it synonymous with *A. palmeri*, and *A. chrysantha* at another.

Nearly sixty years later a graduate student at Arizona State University, Rick DeLamater, reopened the door to this forgotten agave while hiking in the Tonto Basin area. Subsequent exploration eventually uncovered more than ninety sites before it was formally described by Wendy Hodgson and Liz Slauson (1995).[36] The plant is named in honor of Mr. DeLamater, who passed away only a year after its rediscovery.

A. delamateri is the most common of the Arizona pre-Columbian cultigens, in fact far more common than all others combined. To date I have found it at more than 200 sites, and believe there are still more waiting to be discovered. It is glaucous blue and banded, mid-sized, 70–100 cm (27″–40″), with lanceolate

Happy, healthy *A. delamateri* stand growing along flats in an ancient garden setting

A. delamateri keeping company with *Opuntia* east of Globe.

8–10 cm (3″–4″) wide leaves, and a dramatically ascending posture, associated primarily with ancient Salado Native Americans in and around the Roosevelt Lake region. It may have also been cultivated by latter-era Hohokam through their northern and northeastern perimeters before the Salado culture took hold in the region. *A. delamateri* is also found at Sierra Ancha and the Verde Valley, occasionally in exceedingly harsh conditions at elevations as low as 700 m (2,300′). It is also encountered at altitudes as high as 1,500 m (4,900′), where it can tolerate temperatures as low as −10°C (15°F) without issue.

Ancient indigenous Americans employed a different cultivation strategy for *A. delamateri* when compared to *A. murpheyi*. *A. delamateri* was typically grown along alluvial terraces at the moderate elevations of interior chaparral habitat, with no augmentation to manage water resources.[37] Irrigated crops were

Many *A. delamateri* specimens adopt a sharply ascending posture.

A. delamateri in all its glory.

A. delamateri blooms are often visited by favored pollinators such as hummingbirds.

cultivated below. In contrast, *A. murpheyi* was often grown along floodplains and flats, where check dams and rock piles were employed to manage water flow and enhance soil moisture retention, although sites along the northern perimeter were cultivated in much the same manner as *A. delamateri*. The underlying reason for these distinct strategies seems pretty straightforward, since *A. delamateri* was generally cultivated at higher elevations, which benefitted from more rainfall than the broad flats and floodplains below.

Relatively common as it is, the origins of *A. delamateri* remain a mystery. This agave is sterile, producing a few misshapen seed pods toward the top of its inflorescence, which are filled with only white immature seed. Fairly large blooms (50–70 mm) grace long perpendicular panicles atop 5–6 m (16′–20′) stalks through July, and appear quite similar to the red filament version of *A. palmeri*, though buds of *A. delamateri* have a reddish patina *A. palmeri* lacks.

A. delamateri blooms bear a striking resemblance to those of naturally occurring *A. palmeri*.

Salado usually planted this agave in areas surrounded but not immediately populated by naturally occurring *A. chrysantha*. I have on occasion found it in a farm-like setting, the largest of which checking in at an impressive 50 acres, where I have documented 32 remaining *A. delamateri* patches containing perhaps 500 plants. That may not sound like a lot, but in terms of pre-Columbian agave cultigens, it is an absolute bonanza of startling proportion.

The advantage of *A. delamateri* over naturally occurring agaves seems clear. It is less fibrous and sweeter, thus better eating.[38] So much preferred was *A. delamateri* over *A. murpheyi* that *A. murpheyi* domestication seems to have been all but completely abandoned through the Tonto Basin area, despite its advantageous bloom time. While *A. murpheyi* sites are found in ancient Salado haunts, there are at least forty *A. delamateri* sites for each, so a strong preference seems demonstrated. But this was a spiritual lot. Reasons for their preference may have been entirely pragmatic, but might just as well have leaned toward the spiritual. We will never know.

AGAVE PHILLIPSIANA (Grand Canyon agave) is tall, green, and stately. Ascending 70–90 cm (27"–36") lanceolate leaves are moderately banded, 10–12 cm (4"–5") wide, and adorned with an ample supply of somewhat haphazardly oriented marginal spines. Initially believed endemic to the Grand Canyon area by Wendy C. Hodgson (2001),[39] this beautiful agave is named in honor of Arthur Phillips III, author and botanist, who first reported the species to Hodgson in 1994, and monitored the Grand Canyon sites at which it was found. Its actual discovery is credited to Rose Collom, the Grand Canyon's first botanist, in 1938.

It has since been discovered at dozens of sites between 1,000–1,700 m (3,200'–5,600') elevation across central and northern Arizona, including the Prescott National Forest, and red rock country near Sedona, in the heart of Sinagua territory. The incredibly picturesque landscape surrounding Sedona is an archaeological wonderland of the highest order, chocked full of small-to moderate-sized cliff dwellings and numerous pictograph (painted rock art) panels of startling artistry. *A. phillipsiana* seems to have been planted across

A. phillipsiana from the Prescott National Forest.

A. phillipsiana growing alongside the ubiquitous *Yucca baccata*.

A. phillipsiana may be the most striking of the Arizona domesticates.

this area in small garden plots, occasionally nestled into naturally occurring *A. parryi* patches, without obvious interaction since bloom periods do not overlap. *A. verdensis* is also scattered throughout the region.

A few additional *A. phillipsiana* sites are found near Roosevelt Lake in the heart of ancient Salado territory. The origins and cultural associations of Grand Canyon sites seem more elusive. The area was occupied by Ancestral Puebloan until roughly AD 1250, but there is no way to know whether *A. phillipsiana* was cultivated prior to that time or after the Ancestral Puebloan vacated the region and Southern Paiutes filled that void.

Like *A. delamateri*, *A. phillipsiana* is *palmeri*-like and boasts an expansive distribution range, though it is far less common. It may have been as common at one time, but *A. phillipsiana* would not have fared so well at lower elevations once Native American caretakers mysteriously vanished from the region.

Close-up of *A. phillipsiana* flowers.

Flowers are large (60–80 mm) and decidedly bland, with green/cream coloration and modestly ruby-tinged tepals, which, like those of *A. palmeri*, are both erect and calloused. Blooms open mid-August through mid-September along a 3–5 m (10'–16') inflorescence. This beautiful cultigen is seed sterile; in fact stalks are devoid of seed pods entirely, which all drop almost immediately after flowering.

In regard to the sterility of *A. delamateri* and *A. phillipsiana*, agave guru Wendy Hodgson suggests that pre-Columbian Native Americans might have encouraged vegetative reproduction by removing embryonic reproductive structures to prevent flowering. Modern coring techniques attest to the effectiveness of this method in regard to producing offsets and axial branches. Certain mescal producers in Mexico also engage in the practice today. The evolutionary vector from cespitose to sterile seems rather more elusive to this reporter. Perhaps selection played its part until the most prolifically cespitose plants were simply no longer able to set seed.

It is not entirely clear whether *A. phillipsiana* was developed regionally or in Mexico and subsequently sent north as trade fodder. Wendy Hodgson and Andrew Salywon of the DBG have demonstrated a genetic link to *A. palmeri*, but *palmeri* also grows naturally south of the border, well into Sonora, Mexico. It is clear that *A. phillipsiana* is not as tolerant of harsh, dry conditions as *A. delamateri*, and demonstrates greater tolerance toward cold winter conditions. *A. phillipsiana* can manage low desert heat, provided ample water is available, and of course it's extremely hardy, reportedly good down to −20°C (−5°F).

The late bloom time of *A. phillipsiana* represents a substantial harvest time extension over other cultigens and naturally occurring agaves in the region.[40] And not surprisingly, it is sweeter and less fibrous than *A. parryi*, which is common across much of its range. The expansive range of both *A. phillipsiana* and *A. delamateri* and incursion into each other's territory suggest some level of trade between ancient Salado and Sinagua cultures.

A. phillipsiana with emerging bloom stalk.

A. verdensis strutting its stuff.

Pre-Columbian Agave Cultigens

AGAVE VERDENSIS (Sacred Mountain agave) and *AGAVE YAVAPAIENSIS* (Page Springs agave) were not formally described until fairly recently by W. C. Hodgson & A. M. Salywon (2013).[41] At 50–60 cm (20″–24″), *A. verdensis* is a smaller plant than *A. delamateri* and *A. phillipsiana*, and is actually difficult to discern from half-grown *A. delamateri,* with which it associates. Lanceolate to oblanceolate leaves are 6–10 cm (2.4″–4″) wide and absolutely rife with all manner of closely set, strongly recurved needlelike marginal spines. It was first discovered in 1995 near Sedona, and has a very limited range, having not been found outside the Verde Valley, from which it derives its name.

The area in which most *A. verdensis* sites were subsequently discovered is an especially intriguing archaeological region, with Sacred Mountain at the center of an ancient southern Sinagua agricultural community. Canals, check dams, terraces, roasting pits, and rock mulch abound, as well as a multitude of impressive habitation sites and petroglyph (chiseled rock art) panels. One significant petroglyph site was carefully crafted as a solar calendar that marks the winter solstice and other important dates. Many of these sites are fairly obscure, while others are regionally famous, and one has actually been enshrined as a national monument. There are plenty of other agaves about keeping *A. verdensis* company, including *A. delamateri* and *A. phillipsiana*.

Beautiful blue banding of *A. verdensis*.

A. verdensis overlooking the Verde Valley.

A. verdensis with emerging inflorescence.

A. yavapaiensis in favored habitat.

A. *yavapaiensis* was first discovered some fifteen miles distant, just eight years later in 2003. It is rather similar to *A. verdensis* in appearance, but features a more open rosette. Numerous elliptic to linear oblanceolate leaves may be 10 cm longer than those of *A. verdensis*, and a bit narrower at 5–8 cm (2″–3″). The *cabezas* are exceptionally large and dense, weighing 13–18 kg (28–40 lbs.). This agave is named for the county in which it resides, as well as for the Yavapai tribe that may have known and utilized the plant for many years.

These strikingly similar species are nearly indistinguishable until almost half grown. *A. yavapaiensis* grows in small patches with *A. verdensis*, *A. delamateri*, and *A. phillipsiana*, with the vast majority found in two small parcels of the Verde Valley. Both *A. verdensis* and *A. yavapaiensis* cultigens are found at 1,000–1,500 m (3,200′–5,200′) elevation, and each is hardy to –15°C (10°F).

Bloom period for *A. verdensis* commences in mid-June with *A. yavapaiensis* trailing by no more than several days. Flower morphology is similar in both species. Blooms for each are cream/yellow and 40–60 mm, gracing 4–6 m (13′–20′) stalks. Like seed sterile *A. delamateri* and *A. phillipsiana*, these also

Beautiful form of *A. yavapaiensis*.

A. yavapaiensis growing in a grassy Opuntia field.

A. yavapaiensis seems to handle drought conditions better than other regional agaves.

Characteristically tall and narrow *A. yavapaiensis* panicles.

share bloom characteristics such as erect tepals and tepal callouses with *A. palmeri*, but *A. verdensis* and *A. yavapaiensis* actually produce small quantities of seed in the upper sections of their inflorescence. The area in which they grow, however, does not seem especially conducive toward germination and seedling development, so reproduction is almost exclusively asexual.

Details of *A. verdensis* flowers.

Of our five described Arizona pre-Columbian cultigens, these may be the sweetest of all. Cattle do not usually eat agaves, but frequently tear down stalks and consume blooms with great zeal. In one particularly intriguing locale, each and every *A. verdensis* and *verdensis*-like inflorescence is torn down year after year, while *A. chrysantha* stalks remain largely unmolested, demonstrating a clear preference for sweeter *A. verdensis* blooms. I won't bother to mention my annoyance at the hindrance this directs toward further research in the area.

Along with our two described cultigens, I have found small numbers of no less than four distinct unidentified agave types in this area, along with several others whose identity is ambiguous at best. This kind of Agavaceae jambalaya may suggest ancient horticultural trials directed toward development of additional cultigens, and might also suggest a local origin for *A. verdensis* and *A. yavapaiensis*.

On the other hand, agave guru Wendy Hodgson of the Desert Botanical Garden, who formally described these species and now embraces a local origin, initially suspected development in Mexico, in part due to affinities with *A. shrevei* and other regional agaves. *A. shrevei* is of particular interest because, like *A. palmeri*, it also has calloused tepals. That alone makes it suspect when considering the pedigree of our cultigens. Conceding that affinity, I believe it just as likely that *A. shrevei* was brought north as trade fodder and massaged locally to develop our cultigens. A distinct locale some 130 km (80 miles) distant is rife with a perplexing variety of unidentified *A. chrysantha*-like agaves, also laden with apparent *A. shrevei* affinities. Further research is warranted, including a systematic search across areas of northern Mexico to determine whether unknown conspecific or similar cultigens reside in these areas.

Agaves grow below a Sinagua cliff dwelling north of Sedona.

PART III
Notes *from the* Field

CHAPTER 7

The Chase

PRE-COLUMBIAN CULTIGENS ARE ESPECIALLY REWARDing to chase in the field since they're always found in the vicinity of archaeological features including dwellings, terraces, rock piles, roasting pits, rock art, and potsherds, usually in small groups of three to thirty plants, almost always between 750–1,500 m (2,400′–5,000′) elevation along flats or gentle slopes, in what typically present as garden settings. Oft times I'll encounter terraces or an ancient dwelling, then look for associated agaves, or perhaps just the reverse.

Make no mistake, each of these cultigens was exceedingly difficult to find first time out. They do not spread across suitable habitat like naturally occurring agaves, usually remaining confined to the terraces or small garden plots in which their forebears were planted. Also, they are easily confused with *A. chrysantha*, which is almost always lurking in the general area, if not the immediate vicinity. To make things more difficult, some of our cultigens readily hybridize with both *A. chrysantha* and *A. parryi*. With experience it became easier, as I honed my recognition skills and learned to take cues from the topography of each region, including reported archaeological site locales. Over the past several years, I have discovered more than 300 such sites across the state, many of which were previously unknown.

The most challenging cultigens to find are undoubtedly *A. murpheyi* and *A. phillipsiana*. Both tend to blend into their backgrounds more convincingly than bright blue *A. delamateri*, *A. verdensis*, or *A. yavapaiensis*. They are also the most widely scattered. There are at least fifteen *A. phillipsiana* sites spread across the Prescott National Forest at 1,200–1,400 m (3,900′–4,600′) in fairly thick pinyon-juniper woodland, and they are the devil to spot. All are found along slopes, some 5–10 m (16′–32′) below summits, where fully formed bloom stalks never have horizon rather than forest as a backdrop. And *A. phillipsiana* flowers are all but colorless, so nearly as difficult to spot while blooming. I sometimes wonder if ancient caretakers may have intended to camouflage these agaves.

The famed Painted Deer petroglyph near Agua Fria National Monument.

Domesticated agaves are only rarely found within visual range of ancient dwelling sites, almost always planted out at some discreet distance. There are too many unknown variables and possible reasons for this to enumerate, but I oft suspect security as one potentially compelling issue. Many ancient dwellings seem carefully placed with security of paramount concern, and unusual 6 m (20′) agave stalks popping up off season might be too bright a flag to plant directly outside one's front door.

It has always seemed interesting that pre-Columbian cultigens are not usually found within striking distance of naturally occurring agaves. Were *A. delamateri* planted out amongst naturally occurring *A. chrysantha*, the result would be undiluted *A. delamateri*, which does not produce seed, and a regional *A. chrysantha* population bent toward the more desirable characteristics of *A. delamateri*. The impact would not have spread far, since resulting progeny are effectively sterile. Perhaps that sterility is all the explanation needed, especially if pollinators demonstrate a preference for sweeter *A. delamateri* nectar. That said, there are a handful of locales at which both *A. delamateri* and *A. verdensis* converge with *A. chrysantha*. In many cases, such convergence may represent subsequent range expansion by *A. chrysantha*.

With few exceptions, pre-Columbian cultigens found growing in Hohokam and Salado territory are relatively isolated, but Sinagua territory offers an entirely different experience. It isn't unusual to find various combinations of *A. verdensis*, *A. yavapaiensis*, *A. phillipsiana*, and *A. delamateri* growing in the same acreage of the Verde Valley, often separated by no more than a few meters, occasionally in substantial numbers. I have also encountered agaves that appear intermediate between *A. delamateri* and *A. verdensis*, and in another intriguing locale, *A. delamateri*, *A. verdensis*, and *A. yavapaiensis* are growing within visual range of each other along with the northernmost population of *A. toumeyana* known, in what presents as unlikely habitat for *A. toumeyana*. There is no compelling explanation for the presence of *A. toumeyana* in the area, some 60 km (37 miles) north of its nearest conspecific neighbors. It could be naturally occurring, but might just as well have been relocated to utilize as amole.

OTHER PRE-COLUMBIAN CULTIGEN CANDIDATES

It isn't particularly unusual to happen upon isolated groups of unfamiliar agaves with thin, pliable leaves and virtually no vegetative variance, characteristics generally associated with cultigens, rather than naturally occurring agaves. In the absence of recent bloom stalks, all I can do is return during

A couple of sites inhabited by *A. verdensis* and *A. chrysantha* produce intermediate forms.

New, presumed domesticated agave growing in *A. palmeri* territory, southeast of Tucson. I have found this agave at three separate sites.

New domesticated agave, just described as *A. sanpedroensis*.
Photo courtesy of Wendy Hodgson.

Another distinct new agave found at several sites along the San Pedro Watershed.

suspected bloom season year after year until new stalks emerge and flower morphology provides some valuable clues. More times than not, evidence points toward unusual *A. chrysantha* clones, possibly isolated by natural circumstance or selected and planted as cultivated by ancient Native Americans. But once in a while I hit pay dirt, and find what appear to be previously unknown cultigens.

In addition to our five formally described pre-Columbian cultigens, several other contenders have been discovered across the state. I have found some very suspicious Agavaceae characters growing amongst *A. palmeri* east of Tucson, and on and near Perry Mesa in central Arizona, and a handful of additional candidates have been identified by Wendy Hodgson and Andrew Salywon of the Desert Botanical Garden along the San Pedro Watershed.[42]

Agaves at Perry Mesa are particularly enigmatic, as is the prehistory of the area itself. Perry Mesa, contained largely within the 72,000 acre Agua Fria National Monument, is squarely situated between Southern Sinagua and Hohokam territory, and hosts more than 300 archaeological sites including six major settlement regions. Nearly all of these sites are associated with a pre-Columbian culture of unknown origin, apparently distinct from either of its

Suspected domesticated agaves growing in relatively unexplored terrain on Indian reservation

neighbors.[43] I am in complete agreement with the assessment of Wendy Hodgson and Andrew Salywon that some of the agaves here have a strong waft of pre-Columbian cultigen about them, appearing morphologically aligned with *A. verdensis* and *A. yavapaiensis*, and these seem to come in more than one flavor. Many are growing in unlikely habitat, and many appear to be hybridizing. Good luck to the capable DBG researchers attempting to sort this out!

Other potentially new agave taxa dot the landscape across the Verde Valley and Sierra Ancha, with yet another potential candidate north of Lake Pleasant. Lake Pleasant agaves appear similar to depauperate *A. chrysantha* and produce viable seed, but are in the wrong place at the wrong elevation in a possible domestic

New undescribed domesticated agave from Perry Mesa.

Unusual domesticated agave appears intermediate between *A. delamateri* and *A. murpheyi*.

Miniature *A. verdensis* look-alike features entirely distinct flower morphology. Plants are no more than one-third the size of *A. verdensis*.

setting. Unfortunately, there are not many of these plants, such that flower collection and proper analysis may take years before sufficient data are available to discern their identity and formally describe, if warranted. The work continues.

A more recent discovery involves rarely explored territory in the heart of Indian reservation. In addition to a potentially new pre-Columbian cultigen or two, an unfamiliar, fully fertile agave permeates the region in fairly substantial numbers. This agave appears naturally occurring, and shares certain qualities considered regionally unique to *A. phillipsiana* including flower morphology and bloom time. Vegetative characteristics also align well with *A. phillipsiana.* Could this mysterious new agave offer clues regarding the origins of *A. phillipsiana*? Exploring Indian reservation is one thing, but conducting actual research is a tricky business, fraught with complexities of red tape and permit requirements. We'll see what the future holds in this exciting new arena.

Another compelling find resides in another state, just a stone's throw beyond the Arizona border. Here we have perhaps 200 identical clones of an unfamiliar *A. parryi*-like agave spread across a small patch of ground no larger than an acre. These agaves are not infertile, but unusually homogenous and completely isolated, perhaps 40 km (25 miles) from other known agave populations and considerably farther from other known agaves that are even remotely similar. The general area is an absolute treasure trove of archaeological features.

I have more than once found fascinating sites that present as horticultural laboratories, where I imagine ancient Native Americans plying their craft by utilizing some of the same selection and hybridization techniques employed today. In one instance, apparent *A. murpheyi* × *A. chrysantha* and *A. murpheyi* × *A. delamateri* hybrids grow alongside *A. murpheyi* atop the very same knoll, only a few meters apart at 700 m (2,300′) elevation, where no naturally occurring agaves reside. This does not seem a haphazard arrangement, and one can only admire the skill with which ancient Salado horticulturalists might have managed to hybridize agave species with such disparate bloom times.

Another fascinating discovery features four distinct agave types growing nearly side-by-side in the Verde Valley, separated by only a few meters. One is *A. verdensis,* the other three are unknown and undescribed, each with unique flower morphology. One of the unknowns looks for all the world like a miniature version of *A. verdensis*, less than half its size. I imagine Sinagua mad scientists, trying their hand with a little of this and a pinch of that in hopes of producing new cultigens with desired characteristics. Oh, to be a time-traveling fly on the stalk to learn the ways of the ancients!

Evidence of ancient habitation is everywhere if you look closely enough.

Beautiful hybrid appears intermediate between *A. parryi* and *A. verdensis*.

CHAPTER 8

Impact on Naturally Occurring Agaves

THOUGH REPRODUCTIVELY COMPROMISED, OUR five described cultigens all produce pollen and are capable of hybridizing with naturally occurring agaves. *A. delamateri* mixes nucleotides with *A. chrysantha* on a regular basis, and I have more than once searched furiously for *A. delamateri* after encountering obvious *A. chrysantha* × *A. delamateri* introgression to good effect. I have also found an *A. palmeri* population that demonstrates moderate introgression with *A. delamateri*, as well as *A. chrysantha* × *A. murpheyi* hybrids. In the heart of Sinagua territory, several small populations of *A. parryi*-like agaves suggest introgression with *A. verdensis*. A couple of these sites feature both species growing in close proximity, so parentage of intermediates is obvious, but a couple others require a modicum of imagination since the source of *A. parryi* introgression is no longer around. Some of these *A. parryi* hybrids are a sight to behold!

On a related front, it is not inconceivable that fertile, unknown cultigens have gone feral and corrupted a substantial portion of Arizona's *A. chrysantha* population. Large sections of what is generally considered the naturally occurring *A. chrysantha* swarm are not in fact *A. chrysantha* at all. Plants are well within vegetative norms, but flower morphology is dead wrong, more closely aligned with *A. palmeri* and some of our cultigens than with *A. chrysantha*. Some of this disparity might result from introgression with nearby domesticated agaves, but only a small part. And it's all well and good to attempt to describe one or some of these populations as distinct, except that wildly diverse flower morphology on plants growing side-by-side renders adherence to H. S. Gentry's methodology virtually impossible. Two large populations are separated by some 50 odd km (31 miles), a third is 150 km (93 miles) away, and there are others. This is not a small, localized phenomenon—it is expansive.

This conundrum leads me to consider a hypothesis that fertile and unknown pre-Columbian cultigens may have gone feral and integrated with naturally occurring *A. chrysantha* populations. There is little empirical evidence to support such a hypothesis, but all areas where aberrant *A. chrysantha* populations occur are rich in archaeological sites. At the moment, I believe this represents the most plausible explanation. Of course, there are other possibilities. Perhaps this presumed corruption was induced by dramatically larger populations of known cultigens than survive today. Which is to say, much larger than I already imagine. I don't know. In the end, it is a mystery, one that may eventually be addressed by the application of modern next generation sequencing analysis.

One of the more intriguing areas I have studied includes what is undoubtedly the largest single living pre-Columbian cultigen site known in the US, which I refer to as the Fifty Acre Field. In addition to impressive numbers of

Hybrids of *A. chrysantha* and *A. delamateri* are not uncommon in certain areas.

Very unusual hybrid presents as *A. parryi* × *A. yavapaiensis*.

A. delamateri, a great many naturally occurring agaves populate the immediate surrounding area, more or less along a single flattened ridgeline some 8 km (5 miles) long. One might expect a naturally occurring population of *A. chrysantha*, but instead I find a dizzying array of unidentified agaves, the likes of which I have never before seen. The *A.* aff *chrysantha* population nearest the Fifty Acre Field appears heavily laden with *A. delamateri* DNA, which isn't at all surprising. It is surprising that agaves a few kilometers away are distinct from those near the field and more diverse in terms of both vegetative and flower morphology than any other single agave population I have encountered. Many of these plants demonstrate an apparent affinity with *A. shrevei*. It is also conceivable that one or more additional cultigens once populated the region, which add to the unusual characteristics and diversity of these agaves. I suspect ancient Salado of tinkering with the naturally occurring agave population in the area, which is surrounded by a fair number of breathtaking archaeological sites.

Chasing Centuries

In Chapter 5 reference was made to a remote section of west central Arizona where occasional 5–8 m (16′–26′) racemose stalks present on unusual oversized specimens, across some twenty odd kilometers of hillside otherwise populated by *A. mckelveyana*. Small patches of *A. utahensis* are also scattered about the area, radiating away from an otherwise isolated and fairly robust population at the extreme southern edge of its range. Is it remotely possible that ancient Native Americans might have played their horticultural hand back here by distributing or even introducing *A. utahensis* to the region? Or is this just the last gasp of a naturally occurring, but now dramatically receding population? Once again, the area is a regional hotbed of archaeological sites. In the end, this is another fascinating mystery, to which ancient anthropogenic influence might have contributed.

We then have the confounding bloom color issue of naturally occurring *A. palmeri* populations. Most Arizona *A. palmeri* feature green/cream colored flowers with bright red filaments and anthers, very similar to those of *A. delamateri*. But some are the same soft yellow as a couple of our other cultigens, and the distribution of yellow flowers is wildly uneven in regional populations, ranging from 1%–90%. I know of no other agave species that

Impact on Naturally Occurring Agaves

A. palmeri bloom color dimorphism on display.

demonstrates similar bloom color dimorphism. Could this represent another instance of feral pre-Columbian cultigen influence? Once again, I do not know, and frankly, this isn't quite as compelling as the *A. chrysantha* enigma, but whenever I encounter something peculiar to this part of the Agavaceae world, it would be foolish to ignore potential consequences of ancient anthropogenic influence as a contributing factor.

Beyond these palpable issues, the extensive range and influence of pre-Columbian cultigens ultimately lead one to question the degree to which our grasp of the distinction between naturally occurring and domestic agave

One of several unusual agave forms found near the Fifty Acre Field.

represents a false dichotomy. In the end, who is to say which agaves are free of anthropogenic influence and which are not? In 1911, famed botanist William Trelease noted of *A. applanata*, "Long cultivated, but of doubtful origin, and greatly misunderstood because of the difference between juvenile, moderately developed, and mature plants."[44] Anthropogenic influence may have played an extensive role in agave distribution and evolution for thousands of years throughout the Sonoran Desert and beyond. With rare exception, we only suspect those examples of domestication that demonstrate some manner of reproductive impairment. 🌿

CHAPTER 9

Cultivation of Naturally Occurring Agaves

IN ADDITION TO OUR PRE-COLUMBIAN CULTIGENS, I have come across some unusual *A. deserti* var. *simplex* sites of suspicious origin in Hohokam territory in central Arizona. In one locale, a naturally occurring population is found along moderate limestone slopes at elevation, but several additional sites in low-lying flats below seem oddly out of place. Archaeological features are commonly found across the area, and *A. deserti* var. *simplex* is still harvested and consumed today by certain members of the Tohono O'odham Nation.

I have also found quite a few apparently unnatural *A. parryi* populations southeast of Tucson in the Sonoita region,[45] and more yet farther north in the Mazatzal Mountains, Agua Fria National Monument, and the Verde Valley. It seems ancient Hohokam, Salado, and Sinagua were quite fond of it, and planted out *A. parryi* plots in nearby gardens and farms, at lower elevation than that at which it naturally occurs.

A. parryi shares a domestic setting with *A. delamateri* along flats at the base of a mountain at one particularly compelling site in the Mazatzal Mountains, which also hosts a fortified Salado pueblo atop its flattened peak. It is easy to recognize *A. parryi* as unnatural here, apart from the suspicious presence of its known pre-Columbian cultigen companion. Although *A. chrysantha* is plentiful in the area, there is no *A. parryi* sharing the mountaintop with the pueblo, as even that elevation 120 m (400′) above is below the natural range of this montane agave.

Cultivated *A. parryi* sites in the Sonoita region all appear to consist of a single large, green clone of the var. *huachucensis* persuasion, and provide startling presentations of dozens of prominent 80–120 cm (2.6′–4′) agaves sometimes packed so tightly in compact terrain, one risks grave bodily harm by not circumnavigating. Offsetting and seedling development are rampant within the swarm, but the area is apparently too unforgiving to promote reproduction beyond the immediate nurturing environment.

Given the apparent relationship between pre-Columbian Native Americans and agaves, it seems reasonable to suggest that other naturally occurring

Startling *A. parryi* var. *huachucensis* stand looks out of place in grassy flats east of Tucson.

Could these *A. parryi* var. *huachucensis*

agaves, particularly *A. chrysantha* and *A. palmeri*, may have been selected and planted in community garden plots, if not farmed outright, with some regularity, especially before cultigens were developed and widely distributed. One might not notice if *A. palmeri* had been planted here and there in the same Sonoita region that harbors low elevation domestic *A. parryi* populations, for example, since that appears to be within its natural range. Just idle speculation, but that kind of innocuous tinkering might have contributed to the *A. palmeri* bloom color enigma discussed in the previous chapter.

CULTIVATED CACTI?

While out and about, I have on occasion made some rather interesting cactus discoveries as well. This is a trickier business than calling out unnaturally occurring agaves, but a couple of finds jump out as particularly noteworthy. One involves a small but robust *Cylindropuntia versicolor* population northeast of Globe, well out of its natural range. Some of these plants are enormous, and all are found within a kilometer of a hilltop Salado pueblo, most along the very same hill. The small pueblo also boasts several associated *A. delamateri* sites, again all within roughly one kilometer or so. No other *C. versicolor* is found in the entire mountain range. This is not the only putative domestic *C. versicolor* site I have encountered, but it is the most compelling.

Since the area is rife with *C. spinosior*, we are wont to consider advantages of cultivating *C. versicolor* in its stead. While ancients may have consumed cholla fruit, fruits of these two species are essentially indiscernible to the palate, though *C. versicolor* fruits are larger. It seems rather more likely that *C. versicolor* flower buds were consumed and may have been preferred over those of other *Cylindropuntia* species, but it is also conceivable that preference for *C. versicolor* has a ceremonial basis. There are modern Native American examples of both cholla flower bud consumption and ceremonial use from which to draw.

Another unusual find involves an *Opuntia laevis* patch in Salado territory, north of Roosevelt Lake. Indigenous Native Americans not only consumed opuntia fruits, but utilized opuntia mucilage as a binding agent to improve strength and longevity of construction mortars.[46] This extremely rare opuntiad may be kin to another cultigen, *O. ficus-indica*, possibly mixing genes with naturally occurring *O. engelmannii* or *O. chlorotica* in the area. Or it could be a hybrid of local origin, carefully selected by ancient Native Americans for

Cylindropuntia versicolor growing well out of range, adjacent to a Salado hilltop pueblo.

Rare *Opuntia laevis* growing near a Salado pueblo.

its spineless form. Not surprisingly, there is a hilltop pueblo in the immediate vicinity. Either way, it appears unnatural, and may have found its way to modern times via human intervention. Chromosome counts and next generation sequencing analysis would undoubtedly go some distance toward clarifying this intriguing mystery.

The extreme rarity of *O. laevis* may be an unfortunate and relatively recent development. Cattle consume opuntia but are deterred by glochids and spines. It is entirely possible that spineless *O. laevis* may have been eagerly consumed with such zeal that the only examples remaining in the entire region are those inaccessible to cattle along canyon walls and cliff faces.

I also find thick patches of various *Nicotiana* (tobacco) and *Ephedra* (desert tea) species growing in possible domestic settings from time to time. Of course, these plants occur naturally and are not uncommon, so it isn't always easy to discern natural from cultivated populations, but experience suggests that some such finds are more suspicious than others. And just like agaves, other naturally occurring plants found in suspect circumstance often provide clues that lead to encounters with nearby archeological features.

CHAPTER 10

Inevitable Extinction

ALL OF OUR PRE-COLUMBIAN AGAVE CULTIGENS ARE rare and vulnerable, some extremely so, and at high risk of extinction. In the past several years since I first made their acquaintance, some have experienced a precipitous decline. These agaves were not intended to thrive on their own. Crops and gardens were planted and cared for by members of agricultural societies, who as luck would have it, left them to fend for themselves more than five centuries ago. Over that span, I can only imagine the climactic peaks and valleys they must have endured. It is truly remarkable they have persisted as long as they have.

Unfortunately, Arizona has now been enduring prolonged drought conditions for three decades and counting, and the list of sites from which domesticated agaves have succumbed continues to grow. Of course, it isn't just the latest drought. Normal cyclical weather patterns could eventually take down these cultigens. Naturally occurring agave populations wax and wane as fluctuating conditions permit, but there is little to no wax once reproductively compromised cultigens wane beyond a certain point. We lost at least a dozen *A. delamateri* sites alone in just the past two years. Once gone there is no way back.

On a related note, we have both human-caused and lightning-strike wildfires to contend with. The sheer number of wildland fires that strike Arizona each year is staggering. In the ten-year period from 2004–2013, more than 22,000 separate fires burned over three million acres across the state. Without reliable seed production to fall back upon, agave cultigens simply lack the means to recover from the worst of these.

Building and development also represent a serious threat to pre-Columbian cultigens. *A. yavapaiensis* in particular is at grave risk of extinction from development, perhaps during our lifetime. Since *A. yavapaiensis* and other domesticated agaves are considered anthropogenic cultigens, they are exempt from protection under the US Endangered Species Act. The only protection currently available is from the US Forest Service. In fact it would seem remiss at this juncture to fail to acknowledge that although pre-Columbian agave cultigens are not extended any special consideration, all habitat agaves in Arizona are salvage restricted as protected native plants, and cannot be legally removed from habitat without application and approval of special permit requirements.

Better than nothing, but that won't help much when big developers step in with their permits in order.

Of our five formally described cultigens, *A. murpheyi* is the rarest of the rare. I know of only twenty sites scattered across the state. There were twenty-one, but one was destroyed by a decision to clear brush away from an old Tonto Basin ranger station north of Roosevelt Lake just last year. Absolutely tragic, and no small irony, since that is the very organization tasked with its protection.

A handful of domesticated agave populations are afforded some additional protection offered by their locale within a national monument, but even that may be transitory, as the US Antiquities Act itself has recently seemed every bit as threatened as the historic resources it was drafted to protect.

Agave mite lesions on *A. verdensis.*

A further threat is agave mite (aka eriophyid mite). This pernicious microscopic scourge scars leaf surfaces, interferes with reproduction, and even kills its hosts. There has been an alarming proliferation of in situ eriophyid mite infestation across Arizona in recent years, and the Verde Valley seems a hot spot. This unsavory pest has also worked its way into many succulent nurseries and conservatories in Arizona and elsewhere across the Southwest. Several in situ Arizona agave species are affected, though many attacks seem no worse than moderate. I have even found examples of agaves that have apparently rid themselves of mites. Unfortunately, that is not the case with our cultigens, which seem to take a harder hit than naturally occurring agaves. *A. verdensis, A. yavapaiensis,* and *A. phillipsiana* are all under siege to various degrees, and this presents a problem above and beyond that faced by naturally occurring species. Agave mite interferes with vegetative reproduction, which these agaves rely upon almost exclusively. Left unchecked, agave mite, in and of itself, could contribute substantially to the extinction of reproductively challenged agaves with an acutely limited range. Sadly, a familiar *A. verdensis* site succumbed to agave mite just this past year, and several others may face the same fate in the near future.

Finally, if changing climate, land development, and agave mite fail to drive these cultigens to extinction, a lack of genetic diversity might also contribute to their demise. This lack of diversity is hardly surprising from cultigens, some of which rely exclusively upon clonal reproduction.

The in situ loss of these living archaeological relics would be nothing short of tragic. Certainly no less so than the loss or destruction of other archaeological dwellings and artifacts deemed precious. And don't look for help from the Archaeological Resources Protection Act any time soon, as living plants are beyond its mandate. In the end, the only remotely conceivable hope is an umbrella of protection granted under the US Endangered Species Act. But that cannot occur unless disqualifying criteria is reevaluated and adjusted to allow for their inclusion. The future is not bright, boys and girls. We'll need to enjoy these wonderful agaves while we can.

FINAL THOUGHTS

Let's try to end on a positive note, to what extent we are able. The Desert Botanical Garden in Phoenix is keenly interested in maintaining and propagating these species. In fact, DBG herbarium curator and agave guru Wendy Hodgson has actually given voice to the brilliant notion of propagating pre-Columbian agave

A. verdensis planted near the parking lot at Tuzigoot National Monument.

cultigens on a commercial basis for the purpose of mescal production. I don't know if she'll be able to get that particular ball rolling, but it is a wonderful idea!

You can see all five or our pre-Columbian cultigens by paying a visit to the Desert Botanical Garden. All have been there for some time now, and are going about the business of producing offsets and bloom stalks, just like the rest of the DBG's agave collection. You can also see *A. delamateri*, *A. phillipsiana*, and *A. verdensis* at Tuzigoot National Monument in Clarkdale, AZ, site of a partially reconstructed 110-room Sinagua pueblo. At one time, the area immediately surrounding Tuzigoot was home to a fair-sized *A. delamateri* population. Sadly, those plants were bulldozed during a berm construction project, but a cognizant Tuzigoot staff member quickly alerted DBG administration, who sent a crew to salvage what they could. Some of these salvaged agaves were subsequently planted along a narrow strip between the parking lot and visitor center, where *A. verdensis* and *A. phillipsiana* were also added.

As to the future of these plants, *A. murpheyi* is very common landscape fare in the Phoenix area, and should be with us for a long, long time within that context. *A. verdensis* and *A. yavapaiensis* both produce small quantities of seed, so their ex situ survival seems all but assured by the legal collection and distribution of seed to interested parties across the planet. *A. delamateri* and *A. phillipsiana* are at greater risk since each is sterile, but grown under favorable garden conditions, offset prodigiously. And we will all benefit by increasing domesticated agave numbers for distribution to institutions and the public via modern tissue culture techniques. So while their habitat demise may be all but inevitable, with a modicum of care and concern, we could eventually have Arizona pre-Columbian cultigens to enjoy in our gardens and conservatories for generations to come.

    Finally, to anyone visiting sunny Arizona, do yourself a favor and take a day to drive up into the mountains, where time stands still and tall, glorious agave stalks dot the landscape. There is something absolutely magical about gardens planted and nurtured by Gaia herself. And if you listen very closely, and there isn't too much wind, you may hear the whispers of ghosts, still tending agave gardens originally planted centuries ago.

A. *phillipsiana* in red rock country.

PART IV
Addenda

ARIZONA AGAVES	SUBGENUS	STATUS	SIZE	ELEVATION	HARDINESS	BLOOM TIME	PLOIDY
A. ×ajoensis	Littaea × Agave	Hybrid	30–40 cm	950–1050 m	−12°C / 10°F	May	Triploid
A. ×arizonica	Littaea × Agave	Hybrid	30–40 cm	700–1700 m	−12°C / 10°F	Mar–Jun	Diploid
A. chrysantha	Agave		80–180 cm	700–1700 m	−12°C / 10°F	Jun–Jul	Diploid
A. delamateri	Agave	Domesticate	70–110 cm	700–1500 m	−10°C / 15°F	Jul–Aug	Tetraploid
A. deserti var. simplex	Agave		50–100 cm	300–1400 m	−15°C / 5°F	May–Jun	Diploid
A. mckelveyana	Agave		40–50 cm	900–1400 m	−20°C / −5°F	May	Diploid
A. murpheyi	Agave	Domesticate	80–110 cm	500–900 m	−7°C / 20°F	Mar–May	Diploid
A. palmeri	Agave		70–240 cm	1200–1800 m	−10°C / 15°F	Jul–Aug	Diploid
A. parryi var. couesii	Agave		60–120 cm	1500–2400 m	−18°C / 0°F	June	Diploid
A. parryi var. huachucensis	Agave		80–120 cm	1300–2400 m	−18°C / 0°F	June	Diploid
A. parryi var. parryi	Agave		60–120 cm	1300–2400 m	−18°C / 0°F	June	Diploid
A. parviflora	Littaea		12–18 cm	1200–1500 m	−15°C / 5°F	June	Diploid
A. phillipsiana	Agave	Domesticate	70–90 cm	1000–1700 m	−20°C / −5°F	August	Tetraploid
A. schottii	Littaea		30–50 cm	700–1900 m	−12°C / 10°F	June	Tetraploid
A. schottii var. treleasei	Littaea × Agave	Hybrid	35–55 cm	1100–1900 m	−12°C / 10°F	June	Triploid
A. toumeyana var. bella	Littaea		10–15 cm	900–1800 m	−12°C / 10°F	Jun–Nov	Diploid / Tetraploid
A. toumeyana var. toumeyana	Littaea		18–60 cm	900–1800 m	−12°C / 10°F	Jun–Nov	Diploid / Tetraploid
A. utahensis var. utahensis	Littaea × Agave (?)		15–70 cm	900–1800 m	−20°C / −5°F	Mar–May	Diploid
A. utahensis subsp. kaibabensis	Littaea		70–120 cm	900–1800 m	−20°C / −5°F	Mar–May	Diploid
A. verdensis	Agave	Domesticate	50–60 cm	1000–1500 m	−12°C / 10°F	June	Diploid
A. yavapaiensis	Agave	Domesticate	60–70 cm	1000–1500 m	−12°C / 10°F	June	Diploid

GLOSSARY

Acuminate: Tapering to a point.

Aff: Taxonomic lexicon short for Latin "affinis" meaning "similar to". Used after genus, before species, to denote affinity, such as *Agave* aff *chrysantha*.

Agavaceae: A family of plants that include *Agave, Beschorneria, Furcraea, Hesperaloe, Hesperoyucca, Manfreda, Polianthes,* and *Yucca*. Relatively narrow use in this text specifically references the genus *Agave*.

Agave mites: See *Eriophyid mites*.

Aguamiel: Juice drawn from enormous tropical agaves known as maguey.

Alluvial: Relating to clay, silt, sand, gravel, or similar detrital material deposited by running water.

Amole: Any of various plants parts, including roots, of various plants, including certain agaves and yuccas, used as a detergent or soap.

Anasazi: Navajo name for Ancestral Puebloans, which translates to "ancient enemy". Despite widespread use, subjects of the term find its use offensive, preferring the name "Ancestral Puebloan" in its stead.

Anther: The part of a stamen that contains pollen.

Anthropogenic: Originating in and of human activity.

Ball court: An oval, bowl-shaped depression typically 30 m × 15 m, constructed by excavating and piling dirt up in a berm around the perimeter, providing a sloping wall up to 3 m in height. This served as an arena for a type of ball game, with spectators watching from atop the surrounding berm.

Bulbil: A small bulblike structure, especially in the axil of a leaf or at the base of a stem, which may form a new plant.

Cabeza: Spanish for "head". The name given to the leaf base/caudex biomass of the agave, which is roasted and eaten after being uprooted and stripped of leaves.

Cespitose: Forming mats or growing in dense tufts or clumps. In agaves, the word specifically denotes rhizomatous offsetting.

Coprolite: Fossilized dung.

Cultigen: Any plant that is deliberately selected for or altered in cultivation. Cultigens can have names at any of many taxonomic ranks, including those of species, variety, form, and cultivar. The words cultigen and cultivar may be confused with each other. All cultivars are cultigens, because they originate in cultivation, but not all cultigens are cultivars, because some have not been formally distinguished and denominated as cultivars. Though "cultivar" is used informally in our title, these agaves have been formally described as species, rather than cultivars, so are more properly referred to as cultigens.

Decurrence: The state of running downward. In agaves, decurrence is quantifiable and refers to the length which corneous terminal spines continue along leaf edges toward marginal spines.

Deltoid: Reference to leaf and marginal spine shape. Triangular.

Diploid: Referring to two homologous sets of chromosomes.

Eriophyid mites: Microscopic mites that parasitize many kinds of succulent plants, including agaves. Unlike most adult mites that have four pairs of legs, eriophyid mites have only two pairs. They are slow moving and have a distinctive "carrot" shape. They are very host-specific, meaning that species found on one plant type will not usually feed on other plant types.

Ex situ: Not in habitat, typically denotes a greenhouse or garden setting.

F1 hybrid: The first filial generation of offspring from distinctly different parental types.

Filiferous: Producing threads.

Glaucous: Covered with a grayish, bluish, or whitish waxy coating that is easily rubbed off.

Haploid: an organism or gamete cell having only one complete set of chromosomes.

Inflorescence: The complete flower head of a plant including stems, stalks, bracts, and flowers.

In situ: In habitat.

Introgression: The transfer of genetic information from one species to another as a result of hybridization and/or repeated backcrossing.

Lanceolate: Reference to leaf shape. Shaped like the head of a lance; of a narrow oval shape tapering to a point at each end, usually four to six times longer than wide.

Laterals: The main branches of a paniculate inflorescence, also called panicles.

Leptonycteris: Genus of nectar feeding long-nosed bats considered primary pollinators of many agaves and large cacti across the desert southwest. The relationship between agaves and nectar feeding bats is profound, and theoretically responsible for the unusual agave adaptation of monocarpism.

Linear: Reference to leaf shape. Long and narrow, with parallel sides.

Littaea: Subgenus of agaves that includes all members of the genus that produce a spicate or racemose inflorescence. Note that some taxonomists consider the genus *Manfreda* synonymous with *Agave*. These also have a spicate inflorescence, but this group is beyond the consideration of this text.

Maguey: Any of several large agaves of Mexico used as a source of fiber, food, and sap, from which pulque is made.

Meiosis: a type of cell division that results in four daughter cells each with half the number of chromosomes of the parent cell, as in the production of gametes and plant spores.

Monocarpic: A plant or rosette that flowers once, then dies.

Monocot: A flowering plant with an embryo that bears a single cotyledon (seed leaf). *Monocotyledons* constitute the smaller of the two great divisions of flowering plants, and typically have elongated stalkless leaves with parallel veins (e.g., grasses, lilies, palms).

Morphology: The branch of biology that deals with the form of living organisms, and with relationships between their structures.

Oblanceolate: Reference to leaf shape. The opposite of lanceolate with the more pointed end at the base.

Ovate: Reference to leaf shape. Oval, like an egg.

Paniculate: Panicled, arranged in panicles, or like a panicle. A *panicle* is a much-branched inflorescence.

Petroglyph: A form of rock art practiced by ancient Native American cultures. Images are created by removing part of a rock surface by incising, picking, carving, or abrading.

Pictograph: A form of rock art practiced by ancient Native American cultures. Images are created by drawing or painting on a rock wall using naturally occurring paints and dyes.

Platform mound: Large rectangular Hohokam archaeology feature, 1–3 m high, ranging from several hundred to several thousand square meters in area.

Ploidy: The number of complete sets of chromosomes in a cell.

Pueblo: Pueblos are old communities of Native Americans in the southwestern United States. The first Spanish explorers of the Southwest used this term to describe the communities housed in apartment structures built of stone, adobe mud, and other local material.

Quids: Fibrous remnants of plant material, which are chewed, but not usually consumed, such as chewing tobacco.

Racemose: Taking the form of a raceme. Specifically, in reference to the agave inflorescence, intermediate between paniculate and spicate.

Rhizome: Underground stem or shoot.

Sapogenin: Lipophilic triterpene derivative which, similarly to phenol, protects plants against microbes, fungi, and other hostile organisms. Sapogenins are used in the preparation of soaps, detergents, and shampoos.

Spatulate: Reference to leaf shape. Having a broad, rounded end, as a spoon.

Spicate: Unbranched. Arranged in the form of a spike.

Tepal: A segment of the outer whorl in a flower that has no differentiation between petals and sepals.

Tetraploid: Referring to four homologous sets of chromosomes.

Triploid: Referring to three homologous sets of chromosomes. Triploid plants are typically sterile.

ENDNOTES

1. Williams, Collins, et al. 2018
2. Sauer 1965
3. Fish and Fish 2008
4. Fish 2000
5. Elson, Ort, Sheppard, Samples, Anderson, and May 2011
6. Houk 1995
7. Ingram 2008
8. Fish, Fish, Miksicek, and Madsen 1985
9. Doolittle and Neely 2004
10. Nobel 1988
11. Hammerl, Baier, and Reinhard 2015
12. Carrasco 2001
13. Crane and Griffin 1958
14. Sidana, Bikram, and Sharma 2016
15. Choudhuri 2014
16. Hodgson 2001b
17. Ibid.
18. Pinklava and Baker 1985
19. Howell and Roth 1981
20. Gentry 1982
21. Castetter, Bell, and Grove 1938
22. Reveal and Hodgson 2002
23. Castetter, Bell, and Grove 1938
24. Pinklava and Baker 1985
25. Reveal and Hodgson 2002
26. Gentry 1982
27. Fish 2000
28. Hodgson 2013
29. Parker, Trapnell, Hamrick, Hodgson, Parker, and Kuzoff 2007
30. Gibson 1935
31. Nabham 1995
32. Gentry 1982
33. Vokes 1999
34. García-Mendoza, and Chiang 2003
35. Hodgson 2001a
36. Hodgson and Slauson 1995
37. Ibid.
38. Ibid.
39. Hodgson 2001b
40. Hodgson 2013
41. Hodgson and A. M. Salywon 2013
42. Hodgson, Salywon, and Doelle 2018 A brand new formal domesticate description for *Agave sanpedroensis* was published as this book was being rushed to the printer.
43. Ahlstrom and Roberts 1995
44. Trelease 1911
45. Parker, Trapnell, Hamrick, and Hodgson 2014
46. Cardenas, Argüelles-Monal, and Goycoolea 1998

SELECTED BIBLIOGRAPHY

Ahlstrom, Richard V.N. and Heidi Roberts, 1995. *Prehistory of Perry Mesa: The Short-lived Settlement of a Mesa-canyon Complex in Central Arizona, ca. A.D. 1200–1450.* Arizona Archaeological Society 28 (June).

Balls, Edward K., 1962. *Early Uses of California Plants.* California Natural History Guides 10. Los Angeles: University of California Press.

Bean, Lowell John and Katherine Siva Saubel, 1972. *Temalpakh: Cahuilla Indian Knowledge and Usage of Plants.* Banning, CA: Malki Museum Press.

Cardenas, Adriana & Argüelles-Monal, Waldo & Goycoolea, Francisco, 1998. "On the Possible Role of Opuntia ficus-indica Mucilage in Lime Mortar Performance in the Protection of Historical Buildings." *Journal of the Professional Association for Cactus Development.* 3.

Carrasco, David, 2001. *Oxford Encyclopedia of Mesoamerican Cultures: The Civilizations of Mexico and Central America.* Volume 1. Oxford: Oxford University Press.

Castetter, Edward Franklin, Willis Harvey Bell, and Alvin Russell Grove, 1938. "The early utilization and the distribution of agave in the American southwest." *University of New Mexico biological series, v. 5, no. 4, University of New Mexico bulletin, whole no. 335, Ethnobiological studies in the American Southwest,* 6 5, 4.

Choudhuri, Supratim, 2014. *Bioinformatics for Beginners: Genes, Genomes, Molecular Evolution, Databases and Analytical Tools.* Cambridge, MA: Academic Press.

Crane, H.R. and James B. Griffin, 1958. "University of Michigan Radiocarbon Dates III." *Science* 128 Issue 3332 (November), 1117–1123.

Doolittle, William E. and James A Neely, 2004. "The Safford Valley Grids: Prehistoric Cultivation in the Southern Arizona Desert." *Anthropological Papers of the University of Arizona* 70 (November).

Elson, Mark D., Michael H. Ort, Paul R. Sheppard; Terry Samples, Kirk C. Anderson, and Elizabeth M. May, 2011. "A.D. 1064 No more? Re-dating the eruption of Sunset Crater volcano, northern Arizona." Conference Paper presented in the symposium: "Tree-Rings, Environment, and Behavior: The Legacy of Jeffrey S. Dean" at the 76th Annual Meeting of the Society for American Archaeology, March 30–April 3, 2011, Sacramento, California.

K. Fish, Suzanne, 2000. "Hohokam Impacts on Sonoran Desert Environment: Landscape Transformations in the Pre-Columbian Americas," 251–280. In David L. Lentz, ed. *Imperfect Balance: Landsacpe transformations in the Pre-Columbian Americas.* New York: Columbia University Press.

_____, and Paul R. Fish, 2008. *The Hohokam Millennium.* Santa Fe: University of New Mexico Press.

_____, Paul R. Fish, Charles Miksicek, and John Madsen, 1985. "Prehistoric Agave Cultivation in Southern Arizona." *University of Arizona Desert Plants* Volume 7, No. 2: 107–112.

Flint, Richard and Shirley Cushing, 2004. *The Coronado Expedition to Tierra Nueva: The 1540–1542 Route Across the Southwest.* Louisville, CO: University Press of Colorado.

García-Mendoza, Abisaí and Fernando Chiang, 2003. "The confusion of Agave vivipara L. and A. angustifolia Haw., two distinct taxa." *Brittonia*, Vol. 55, No. 1 (January), 82–87.

Gentry, H. S., 1982. *Agaves of Continental North America.* Tucson: University of Arizona Press.

Gibson, Frederick, 1935. "Agave murpheyi, A New Species." *Contributions from Boyce Thompson Institute* 7: No, 1: 83–85.

Gregonis, Linda M. and Karl J Reinhard, 1979. *Hohokam Indians of the Tucson Basin.* Tucson: University of Arizona Press.

Griffin, M. Patrick, 2004. "The origins of an important cactus crop, *Opuntia ficus-indica* (*Cactaceae*): new molecular evidence." *American Journal of Botany* (November), 1915–21.

Griffiths, Anthony J.F., Jeffrey H. Miller, David T. Suzuki, et al, 2000. *An Introduction to Genetic Analysis.* 7th edition. New York: W. H. Freeman.

Hammerl, Emily E., Melissa A. Baier, and Karl J. Reinhard, 2015. "Agave Chewing and Dental Wear: Evidence from Quids." *PLoS ONE, 10*(7), e0133710.

Hodgson, Wendy C., 2001a. *Food Plants of the Sonoran Desert.* Tucson: University of Arizona Press.

_____, 2001b; "Taxonomic Novelties in American Agave (*Agavaceae*)." *Novon 11*: 401–416.

_____, 2013. "Pre-Columbian Agaves: Living Plants Linking an Ancient Past in Arizona." In Marsha Quinlan and Dana Lepofsky. *Explorations in Ethnobiology: the Legacy of Amadeo Rea*. Society of Ethnobiology: 101–131.

_____ and A. M. Salywon, 2013. "Two new Agave species (Agavaceae) from central Arizona and their putative pre-Columbian domesticated origins." *Brittonia* (March) 5–15.

_____ and Liz Slauson, 1995. "*Agave delamateri* (Agavaceae) and its role in the subsistence patterns of pre-Columbian cultures in Arizona." *Haseltonia* 3: 130–140.

Hodgson, Wendy C., Andrew M. Salywon, and William H. Doelle, 2018; "Hohokam Lost Crop Found: A New Agave (Agavaceae) Species Only Known from Large-scale pre-Columbian Agricultural Fields in Southern Arizona"; *Systematic Botany*; (Aug).

Houk, Rose, 1995. *Salado: Prehistoric Cultures of the Southwest.* Tucson: Western National Parks Association.

Howell, D. J. and Barbara Schropfer Roth, 1981. "Sexual Reproduction in Agaves: The Benefits of Bats; The Cost of Semelparous Advertising." *Ecology* (February) 1–7.

Ingram, Scott E., 2008. "Streamflow and Population Change in the Lower Salt River Valley of Central Arizona, ca. A.D. 775 to 1450." *American Antiquity* 73, Issue 1 (January), 136–165.

Nabhan, Gary Paul, 1995. "Finding the Hidden Garden." *Journal of the Southwest* 37 (August) 401–413.

Nobel, Park S., 1988. *Environmental Biology of Agaves and Cacti.* Cambridge: Cambridge University Press.

Parker, Kathleen C., Dorset W. Trapnell, J. L. Hamrick, and Wendy C. Hodgson, 2014. "Genetic and morphological contrasts between wild and anthropogenic populations of *Agave parryi* var. *huachucensis* in south-eastern Arizona." *Annals of Botany* 113 (May) 939–952.

_____, Dorset W. Trapnell, J. L. Hamrick, Wendy C. Hodgson, Albert J. Parker, and Robert K. Kuzoff, 2007; "Genetic Consequences of Pre-Columbian Cultivation for *Agave murpheyi* and *A. delamateri* (*Agavaceae*)," *American Journal of Botany* 94:1479–1490.

Pinklava, D. J. and M. A. Baker, 1985. "Chromosome and Hybridization Studies of Agave," *University of Arizona Desert Plants* Volume 7, No. 2; 93–100.

Reid, Jefferson and Stephanie Whittlesey, 1997. *The Archeology of Ancient Arizona*. Tucson: University of Arizona Press.

Reveal, J. L. and W. C. Hodgson, 2002; "Agave Linnaeus." In: Flora of North America Editorial Committee, eds. *Flora of North America North of Mexico*. New York and Oxford. Volume 26, pp. 442–461.

Sauer, Carl O., 1965. "Cultural factors in plant domestication in the new world." *Euphytica*, Volume 14, No. 3: 301–306.

Sidana, Jasmeen, Bikram Singh, and Om P. Sharma, 2016. "Saponins of Agave: Chemistry and Bioactivity." *Phytochemistry* 130 (October) 22–46.

Starr, Greg, 2012. *Agaves: Living Sculptures for Landscapes and Containers*. Portland, OR: Timber Press.

Trelease, William, 1911. "Revision of the Agaves of the Group Applanatae." *Missouri Botanical Garden Annual Report*. Volume 1911 (October) 85–97.

Vokes, Arthur W., 1999. "The Ancient Use of Seashells in Arizona and Beyond." *Old Pueblo Archeology* (September) 1–6.

Williams, Thomas J., Michael B. Collins, Kathleen Rodrigues, William Jack Rink, Nancy Velchoff, Amanda Keen-Zebert, Anastasia Gilmer, Charles D. Frederick, Sergio J. Ayala, and Elton R. Prewitt, 2018." Evidence of an early projectile point technology in North America at the Gault Site, Texas, USA." *Science Advances*, Vol 4, No 7.

INDEX

A

Agave ×*ajoensis*, *38*, 39, *40*, 41, 79
Agave angustifolia, 98
Agave applanata, 136
Agave ×*arizonica*, *42*, 43, *44*, 45, 79
Agave chrysantha, ix, 16, *34*, *37*, 43, 45, *46*, 47, *48*, *49*, 61, 62, 63, 67, 69, 76, 79, 81, 83, 93, 99, 104, 117, 121, 122, *123*, 124, 126, 128, 131, *132*, 133, 135, 137, 140
Agave couesii, 67
Agave delamateri, x, *14–15*, 16, 36, 48, 57, 61, 63, 69, 92, 93, 99, *100*, 101, *102*, *103*, 104, 107, 108, 111, 114, 121, 122, *127*, 128, 131, *132*, 133, 134, 137, 140, 143, 146, 147
Agave deserti, 51, 52, 54
Agave deserti var. *simplex*, *34*, 39, *40*, 41, *50*, 51, 52, *53*, 54, *55*, 57, 76, 137
Agave fourcroydes, 17, 26, 27, 98
Agave mckelveyana, 48, *56*, 57, *58*, 59, 67, 69, 89, 134
Agave mite, 41, *144*, 145, 153
Agave murpheyi, 6, *7*, *10*, *11*, 21, 24, 36, 48, *91*, 92, 93, *94*, 95, 96, *97*, 98, 101, 103, 104, 121, *127*, 128, 131, 144, 146
Agave neglecta, 48
Agave palmeri, 27, *34*, *37*, 47, *60*, 61, *62*, 63, *64–65*, 69, 73, 75, 76, 93, 99, 103, *104*, 107, 108, 117, *123*, 124, 131, 134, *135*, 140
Agave parrasana, 69
Agave parryi, ix, 16, *37*, 47, 48, 51, 52, *53*, 54, 59, 63, 67, 68, 69, *70–71*, 75, 76, 89, 93, 107, 108, 121, 128, *130*, 131, *133*, 137, 140
Agave parryi subsp. *neomexicana*, 67
Agave parryi var. *couesii*, 68
Agave parryi var. *huachucensis*, 67, 137, *138–139*, 140
Agave parryi var. *parryi*, 66, 68

Agave parviflora, 63, *72*, 73, 75
Agave phillipsiana, 36, 59, 69, 85, *92*, 93, *105*, *106*, 107, 108, *109*, 111, 114, 121, 122, 128, 145, 146, 147, *148–149*
Agave repanda, 99
Agave sanpedroensis, *124*
Agave schottii, vii, 27, 39, 48, 69, *74*, 75, *76*, *77*, 79, 83
Agave schottii var. *treleasei*, 39, *78*, 79
Agave shrevei, 117, 133
Agave simplex, 54
Agave sisalana, 26, 27
Agave tequilana, 62, 98
Agave toumeyana, ix, 16, 27, *37*, 43, 45, 48, 73, 75, *80*, 81, *82*, 83, 122
Agave toumeyana var. *bella*, 45, *82*, 83
Agave ×*treleasei*, 79
Agave utahensis, 43, 57, 59, 85, *86*, 89, *90*
Agave utahensis subsp. *kaibabensis*, *84*, 85, 87
Agave utahensis var. *eborispina*, 85, *88*
Agave utahensis var. *nevadensis*, 85, 89
Agave verdensis, *11*, 36, 48, 92, 107, *110*, 111, *112*, 114, 117, 121, 122, *123*, 126, *127*, 128, *130*, 131, *144*, 145, 146, 147
Agave yavapaiensis, 36, *92*, 111, *113*, 114, *115*, *116*, 117, 121, 122, 126, *133*, 143, 145, 147
Agua Fria National Monument, *120*, 124, 137
amole, 27, 75, 79, 83, 122, 153
Anasazi. *See* Ancestral Publoan
Ancestral Puebloan, xi, 8, 9, 12, 107, 153
Archaic Tradition, 3
Archilochus alexandri, 48
Aztec, 25

B

Besh Ba Gowah, 16
Black-chinned hummingbird, *48*

Index

C
Casa Grande Ruins National Monument, 6
Cylindropuntia spinosior, 141
Cylindropuntia versicolor, 140, 141

D
Desert Botanical Garden, xi, 36, 45, 47, 54, 63, 98, 108, 117, 124, 126, 145, 146
diploid, 35, 36, 39, 41, 45, 83

E
Ephedra, 142
eriophyid mite. *See* Agave mite

F
Ferocactus wislizeni, 76
Fouquieria splendens, 51

G
Gentry, Howard Scott, 32, 43, 51, 52, 57, 67, 85, 89, 93, 95, 99, 131

H
Hodgson, Wendy, xi, 36, 39, 47, 98, 99, 105, 108, 111, 117, 124, 126, 145, 157, 158
Hohokam, xiii, 5, 6, *7*, 8, 9, *11*, 12, 16, *18–19*, 20, 24, *28*, 48, 51, 69, 75, *91*, *94*, 95, 98, 101, 122, 124, 137, 157
Hopi, 16

L
Leptonycteris bats, 61, 62

M
Marana rock mulch farm, 20, 24
Mayahuel, 25
Mayans, 17, 27
McKelvey, Susan D., 57, 99
Mogollon, xi, 8, 12
Montezuma Castle, 9
Montezuma Well, 9

N
Navajo, 8, 153
Nicotiana, 142

O
Opuntia, 100, 115
Opuntia chlorotica, 141
Opuntia engelmannii, 141
Opuntia ficus-indica, 141
Opuntia laevis, 141, 142

P
Palatki Heritage Site, 9
ploidy, 35, 36, 39, 45, 83
Pueblo Grande Museum, 6
pulque, 25, 154

S
Safford Valley, 21, *22–23*, 24
Salado, xiii, 12, *13*, *14–15*, 16, 20, 24, 48, 63, 69, 75, 83, 101, 104, 107, 108, 122, 128, 133, 137, 140, 141, *142*
Sinagua, *xii*, xiv, 8, 9, *10*, *11*, 16, 20, 69, *93*, 105, 108, 111, *118*, 122, 124, 128, 131, 137, 146

T
tepal callouses, 34, 61, 117
tetraploid, 35, 39, 45, 83
Tohono O'odham, 16, 51, 137
Tonto Creek, 12, 16, 96
Tonto National Monument, 16
Trelease, William, 99, 136
triploid, 35, 39, 45, 79
Tuzigoot National Monument, 9, 146

V
V Bar V Ranch, 9
Verde Valley, 8, 9, 101, 111, *112*, 114, 122, 126, 128, 137, 145

W
Wupatki National Monument, 9

Y
Yucca baccata, 106
Yucca elata, 27

Z
Zuni, 16

RON PARKER is an outdoorsman, xeric plant enthusiast, and amateur botanist who spends half his time gardening and the other half exploring natural habitat across Arizona and neighboring states, primarily chasing agaves and archaeological sites. He has been studying agave populations in Arizona for many years, and has been out in the field with renowned botanists and regional archaeologists. When not under the open sky, Ron maintains the well-known xeric plant discussion forum, Agaveville.org, an impressive online repository for information on agaves and other succulent plants.